Fire Insurance Risks and Claims

"Is this a class B. fire ?"

Fire Insurance Risks and Claims

A Guide for Construction Professionals

A.J. Oakes *FRICS, FCIArb*

Blackwell Science

Editorial Offices:
Osney Mead, Oxford OX2 0EL
25 John Street, London WC1N 2BL
23 Ainslie Place, Edinburgh EH3 6AJ
350 Main Street, Malden
MA 02148 5018, USA
54 University Street, Carlton
Victoria 3053, Australia

Other Editorial Offices:

Blackwell Wissenschafts-Verlag GmbH
Kurfürstendamm 57
10707 Berlin, Germany

Zehetnergasse 6
A-1140 Wien
Austria

First published 1997

Set in 11/14pt Bembo
by DP Photosetting, Aylesbury, Bucks
Printed and bound in Great Britain
by Hartnolls Ltd, Bodmin, Cornwall

DISTRIBUTORS
Marston Book Services Ltd
PO Box 269
Abingdon
Oxon OX14 4YN
(*Orders:* Tel: 01235 465500
Fax: 01235 465555)

USA
Blackwell Science, Inc.
Commerce Place
350 Main Street
Malden, MA 02148 5018
(*Orders:* Tel: 800 759 6102
617 388 8250
Fax: 617 388 8255)

Canada
Copp Clark Professional
200 Adelaide Street, West, 3rd Floor
Toronto, Ontario M5H 1W7
(*Orders:* Tel: 416 597-1616
800 815-9417
Fax: 416 597-1617)

Australia
Blackwell Science Pty Ltd
54 University Street
Carlton, Victoria 3053
(*Orders:* Tel: 03 9347-0300
Fax: 03 9347-5001)

A catalogue record for this title is available from the British Library

ISBN 0-632-04017-3

Contents

Preface

The risk of fire is endemic in design and construction of buildings, constant in management of buildings, recurs in the refurbishment of and alterations to buildings, and is ever present in all existing buildings. Building regulations will limit the risk if properly understood and adhered to, but these are mainly directed towards preventing damage to health, safety and welfare of persons rather than the fabric of buildings. The need for special attention in the matter of insurance against material damage to buildings becomes the responsibility of the construction consultant throughout the design, management, alteration, refurbishment and the entire life of a building.

Fire damage to buildings in the UK amounts to approximately £1 billion annually. The majority of this money is paid by the insurance companies to the construction industry and associated professions. Most of the causes and extent of fire in buildings arise from human error, either in building design, maintenance or management, or in fire fighting, either by occupants at the incident of the fire or by professional fire fighters. This book confines itself to principles and points the way to further reading in the attempt to reduce the effect of human error. In addition, it aims to assist the professional consultants, not only to understand the effect of fire but to put their clients in the best possible situation to prevent that damage and to safeguard them financially should fire damage occur.

Within this framework the book deals comprehensively with understanding fire, the effect of fire, measures of prevention, occupation, fire certificates and management, and in detail provides guidance to the risks and claims on the whole insurance situation of sums to be insured, valuations, indemnity, disclosure, claims and reinstatement. Practical examples are given to assist the construction professionals, together with legal cases to illustrate principles and these in themselves make interesting reading. Once the book is in the reader's bookcase, it will be immensely valuable for reference. In this respect, the writer has duplicated certain matters in a very limited way, so that each chapter can be read on its own without too much reference to other parts of the book.

This book is written in the style of the Aqua Group, the author being a founder member. The Aqua Group currently has *Pre-Contract Practice* (eighth edition), *Contract Administration* (eighth edition) and *Tenders and Contracts for Building* (second edition) published by Blackwell Science.

Acknowledgements

The author and publishers would like to thank the London Fire and Civil Defence Authority for permission to reproduce a typical *Fire Certificate* as shown in Appendix 3.

The author has been fortunate to receive advice and contributions of text from the following: James Williams DA Edin., FRIBA, in respect of Chapters 2, 3 and 4; Roy Young of the Loss Prevention Council in respect of design and insurance requirements; Michael Buckle FCII, MIRM, for his comments on insurance generally and insurance broking; Robert Oakes MA (Cantab), for his direction to law cases suitable to illustrate the text; and finally, but by no means least, Brian Bagnall. Brian was a founder member of the Aqua Group and his delightful sketches help so much to put humour and fun into a serious and professional text. Many thanks to them all.

Schedule of Law Cases

The following law cases are given to illustrate the text in the book. Before relating comments in the book to new cases the reader is advised to refer matters to solicitors fully versed in construction law and insurance.

It is a feature of litigation that most competent counsel are able to find authorities on a particular issue which, at least on first reading, appear to conflict. Therefore the cases quoted should not be taken as statements of principle without further advice being sought by specialist lawyers.

Chapter 2

Austin v. *Drewe* (1816) 6 Taunt 36; 2 Marsh 130; 128 ER 1104.

Everett v. *London Assurance Co.* (1865) 19 CBNS 126; 6 New Rep. 234; 34 LJCP 299; 11 Jur NS 546; 13 WR 862; 144 ER 734.

Chapter 3

Anns v. *Merton London Borough Council* (1978) AC 728; 5 BLR 1.

Murphy v. *Brentwood District Council* [1991] 1 AC 398; 50 BLR 1.

Tesco Stores Ltd v. *Wards Construction (Investment) Ltd* (1995) 76 BLR 94.

Chapter 8

Exchange Theatre Ltd v. *Iron Trades Mutual Insurance Co. Ltd* (1983) 1 Lloyd's Rep. 674.

Chapter 9

Dominion Mosaics Co. Ltd v. *Trafalgar Trucking Co.* [1990] 2 All ER 246.

Symington v. *Union Insurance Society of Canton* (1928) 97 LJKB 546.

Yorkshire Water Services Ltd v. *Sun Alliance and London Insurances Ltd* (not formally reported).

Chapter 10

Pan Atlantic Insurance Co. Ltd v. *Pine Top Insurance Co. Ltd* [1995] 1 AC 501.

Chapter 11

National Trust v. *Haden Young Ltd* (1994) 41 Con LR 112; 72 BLR 1.

Chapter 12

Reynolds v. *Phoenix Assurance Co. Ltd* [1978] 2 Lloyd's Rep. 22, 440.

Pleasurama Ltd v. *Sun Alliance Insurance Co. Ltd* [1979] 1 Lloyd's Rep. 389.

Leppard v. *Excess Insurance Co. Ltd* [1979] 1 WLR 512.

Digital Equipment Co. Ltd v. *Hampshire County Council* [1996] 4 All ER 336.

Church of Jesus Christ of Latter Day Saints (Great Britain) v. *Yorkshire Fire and Civil Defence Authority* (1996) QBD Times May 9 1996.

Caparo Industries plc v. *Dickman* [1990] 2 AC 605.

FIRE

Chapter 1
Introduction

Very often a designer is uneasily aware of his limited knowledge of fire hazards and his heavy responsibility over matters of safety. But such is the proliferation of documents on the many aspects of the subject that a form of defensive amnesia sets in and he chooses to push on regardless, relying on those in control to step in with guidance and solutions to his design problems before any real damage can be done.

The consequences of fire seem to be the domain of very few specialists, mostly insurance experts and surveyors acting for the insurance institutions. Faced with instructions to act on behalf of a building owner, to analyse the damage, advise on reinstatement and give guidance on optimising insurance benefit, many construction professionals become embarrassingly aware of their shortcomings. What they will undoubtedly learn is that a better understanding of fire and insurance before the event would have made their task very much easier than after the damage has occurred.

Legislation, Codes of Practice and the rules governing insurance are constantly changing. Statutory requirements actually differ from area to area. London and many of the large administrative areas have their own bylaws; Scotland's regulations differ from those in England. Even to fire prevention officers, revision to Acts of Parliament can become bewildering. A definitive volume encompassing the entire subject would be impenetrable and impractical. Like the global issue of monthly *Notices to Mariners* advising on the constant change in navigational waters, an encyclopaedia of information on fire would require a constant issue of errata and addenda.

The litigious attitude adopted by many associated with buildings, which arises from design, construction, use and insurance, is yet a further risk to be defended or placed in the hands of expert witnesses. Detailed case law rests in the capable hands of solicitors who specialise in fire and construction. However, principles can usually be simply stated in order to guide the construction consultant along the right course. A single volume describing in simple terms the basic matters about fire will be an essential guide to construction professionals in their knowledge of risks and claims in fire insurance. This book attempts to fulfil that role.

'The eternal triangle'

Chapter 2
Understanding Fire

The nature of fire

The loss of life and property by fire most commonly occurs due to ignorance about the nature of fire and its behaviour, in circumstances that normally appear perfectly harmless. If a successful campaign is to be conducted against fire and its dreadful consequences, then the adversary must be fully understood. It is important never to forget that prevention is better than cure.

To occur at all, fire relies on the presence of three ingredients – fire's own eternal triangle (Fig. 2.1). The parties to the triangle are oxygen, a combustible material and an appropriate level of heat.

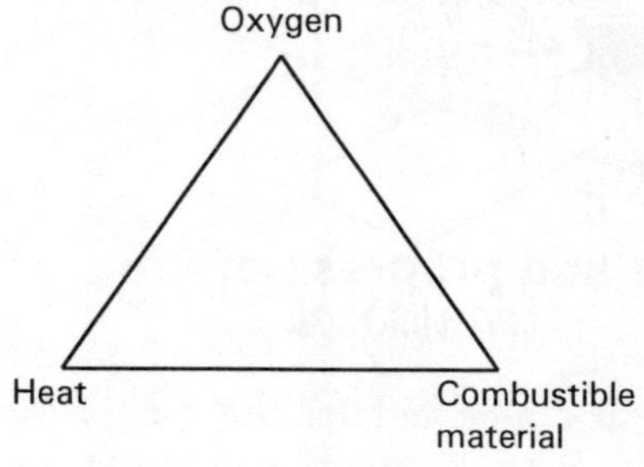

Fig. 2.1 Fire's eternal triangle.

Provided that there exists a combination of these three in the appropriate relationship, combustion will occur. The balance of any two may be correct and it requires only the introduction of the third for the process to begin.

The triangle at work

For very practical reasons the designer must be wise to the workings of the triangle. These are best demonstrated by examples that must seem simple but may be surprising.

Most people are aware that combustion can be prevented by dousing with water. This has the effect, not of cutting off the oxygen supply as might be supposed, but of reducing the local temperature to a level below which combustion will not continue. Water hosed into a raging fire will merely evaporate and sustain the oxygen supply. Water sprayed around the perimeter of the fire will prevent its spread. However, water sprayed onto a hay barn or granary can promote organic decomposition, causing the temperature to rise and a build-up of heat that cannot escape. The consequence would be a fire. One is taught that a small fire may be suffocated by throwing a blanket over it. However, it is not necessarily appreciated that this particular action is only successful if suffocation of the fire continues with a cooling off period. Water on a deep fat fire causes momentary loss of oxygen but the fat, at a very high temperature, instantly evaporates the water and scatters fine droplets which explosively re-ignite. On the other hand, foam or an asbestos blanket is not thrown off by the boiling fat, which gradually cools. In a different context a fire can be starved by isolation from the heat source, for example when the power supply is cut off from an overloaded wiring system.

Separation of one or other of the members of the triangle is the aim of fire fighting.

The evidence and progress of fire

It can be said that the very process of combustion generates the ideal conditions for its own development into full active fire. In fact it is only in something like striking a match that there is an immediate and apparent fire in the form of flame. The usual evidence of combustion in its various stages is as follows:

- smouldering
- smoke

- incandescence (glowing heat)
- flaming.

Figure 2.2 shows the normal progress of combustion in all its phases in a single place and under ideal circumstances, i.e. when there is an appropriate fuel, a rising and maintained level of heat and a steady supply of oxygen – all the circumstances to be found in an airy building filled with combustible furniture but having no fire protection. When a fire rages uninhibited through a building the physical process illustrated below progresses through every part of the fabric, rather like a wave unrolling along a beach and then gradually receding. The development of fire is dependent upon the continuing supply of oxygen. In the first place there is nearly always sufficient oxygen present for some degree of combustion to be initiated even if it is only within the fabric of the fuel material itself. With adequate oxygen present, a non-flaming fire will develop rapidly, emitting heat and building up the temperature required to make the transition to flaming fire. In a poorly ventilated space in which combustion is rapidly consuming the oxygen in the atmosphere, it only requires a depletion by between 10 and 15% for oxygen starvation to occur.

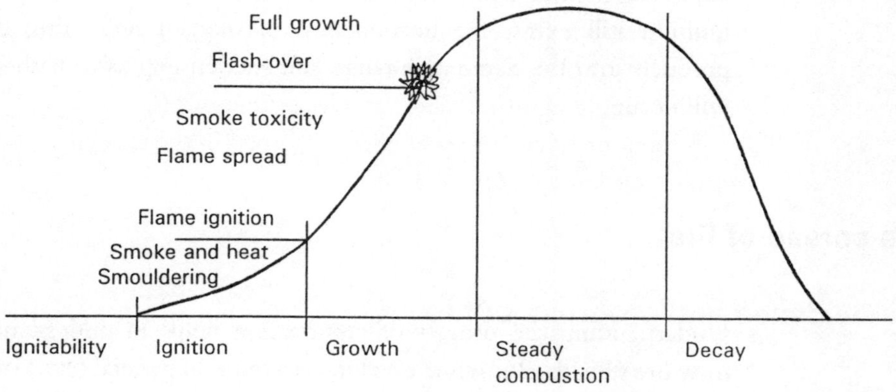

Fig. 2.2 The normal progress of combustion – the wave of fire.

Flash-over and smoke explosion

Because of the reliance of a fire upon oxygen supply, partial combustion with the production of toxic smoke is a phenomenon generally associated with small and often concealed spaces such as cupboards, ducts, ceiling voids and rooms of domestic scale. Where partial combustion occurs in large spaces such as warehouses, factories or shopping malls, it is probably because the combustion material has not achieved the temperature level at which the transition can be made from developing to flaming fire. In either case this transition can be made suddenly and over a large area – this is known as a flash-over.

In broad terms, flash-over occurs when an ordinarily developing fire, or in certain circumstances a smouldering fire, reaches a level of activity at which previously released volatile combustion gases will suddenly ignite and produce flaming across the surfaces of adjoining pre-heated combustible materials.

This very rapid spread of flame differs somewhat from another form of flash-over known as smoke explosion, which occurs when there is oxygen starvation resulting in only partial combustion or a smouldering fire at a deceptively low temperature. The fire leads to the formation of highly volatile vapours which, mixed with a mist of low volatility materials, is regarded as smoke and in the vitiated atmosphere may be of sufficient concentration to be above the upper explodeable limit without actually igniting. Provided the source of ignition still exists, the introduction of oxygen will bring these products into the flammable range and sudden explosive flash-over will occur.

The spread of fire

Understanding the process of combustion helps in understanding how fire spreads. But while one tends to think in general terms of fire as the hazard and destructive force, it is more accurate and useful to think in the specific terms of heat, smoke and flame; each in their own way is harmful to materials, injurious to life and needs to be restricted and fought.

Those who design and manage buildings must constantly remember the three basic media in which smoke, heat and flames are transmitted:

- building contents, e.g. furniture, fittings and fabrics
- building fabric, e.g. finishes, substrate and structure
- building spaces, e.g. concealed spaces, circulation spaces, communicating spaces – any space where smoke can gather.

Building contents

There is no doubt that the major fuel for fires is provided by the building's contents; statistics confirm that it is with the contents that most fires start. While designers are usually aware that the furniture and fittings they specify need to be flame resistant, those responsible for management of buildings are bound to have great difficulty in exercising control over users – over what they bring into the building, the changes they make and the way they behave. The main problem with furniture and fittings lies in the intense burning characteristics that they possess and the rapidity with which combustion products, smoke and toxic gases are released. It is these that are the cause of the majority of fatalities from fire in the United Kingdom, particularly in the home. Curtains and some types of carpet also contribute to the rapid spread of flame unless treated, for they have a more direct contact with building finishes and structures, and readily become involved in the flash-over situation. The greatest risk, however, lies in bedding and upholstery.

Building fabric

The vulnerability to ignition of a material depends on its chemical composition and its surface exposure to air. For example, wood shavings will ignite and burn more readily than the same weight of a similar wood in a single block. It follows that a combustible material used as a finish is less at risk if it is in a dense smooth form, rather than cellular or fibrous. Many popular natural finishes enjoy no such benefit. Man-made finishes can be designed to cope with this situation. However, some man-made products pose the greatest risk of spread of flame because of their chemical composition, for example plastics.

The performance of the materials of which roofs are constructed is important. The inner linings of the roof are those most directly exposed to the heat, smoke and flames of any fire within. Ceiling linings may fall before igniting, exposing the structure behind, which in its turn may fall while burning, spreading flames to furniture, fabrics and floor finishes below. On the other hand, should the roof linings be well secured and noncombustible, or should there be no lining at all, then the roof may confine heat and combustion products, assisting the spread of flame within the building. Roof lights are a particular source of danger as those that melt can produce burning droplets, which ignite materials below. In addition, the collapse can cause the release of volatile gases, which immediately explode in the air and spread flame across the top surface of the roof.

The outer face of the roof is vulnerable, not only to fire breaking through from below, but to fire of burning debris from an adjoining building or a distant part of the same building. Surface spread of flame on the top face of a roof is unrestricted by lack of air. Ignited material such as asphalt on a flat roof will not melt and run away but, sustained by the build up of heat underneath, will eventually burn, disgorging combustion products into the atmosphere and exposing the surrounding areas to smoke hazards. Bitumen burns easily. Either might be disastrous if the roof forms an escape route, particularly on a multistorey building where rescue could depend on the use of helicopters. Collapse of the roof membrane will cause the burning surface materials to become additional fuel for the fire below, the accumulated heat of which might cause smoke explosion. The total effect will be such a build up of heat that the primary structure of the building is endangered, be it brick, steel or concrete.

Of all the constraints, those associated with fire are probably the least considered when building elevations are being designed. Yet it can be the outer face of the building that offers the most direct passage for flames from one part of a building to another. Two factors are involved. The first is the response of the cladding materials to heat and flame; the second is the ability of flames to reach from one opening to another, regardless of the combustion characteristics of the intervening materials.

It is ironic that, by tradition, low buildings are generally faced in safe and noncombustible materials such as brick and stone (timber would not be used where spread of flame might be a critical factor) while modern construction methods used for multi-storey buildings foster the use of cladding materials that are liable to burn or fail in their mechanical performance when subjected to high temperatures. In most incidents fire reaches the face of a building from its own interior by shattering the glass in windows. The window openings then behave as horizontal flue outlets disgorging heat and combustion products up the face of the building. The sudden and plentiful supply of oxygen can, as has already been described, cause sudden eruptions of flaming combustion across the face of materials above and (to some degree) to either side of the opening. It is at this point that the response of the materials involved becomes relevant.

Glass in adjoining windows, panels and curtain wall systems will shatter, scattering lethal debris on people below, leaving insulation exposed to add fuel to the fire while glazing mullions and transoms buckle under the heat. In the same circumstances, glass fibre reinforced plastic panels may distort, melt and (at high temperatures) burn with ferocious activity, spreading flame across the face of the building. Developments in these materials have diminished the risk of surface spread of flame, while glass fibre reinforcing and the use of good mechanical fixings prolong the resistance to collapse. Nevertheless, polymer materials are essentially combustible and the fearful lesson of the Summerland Leisure Complex in 1973 remains with us by the circumstances given in Chapter 7.

Concrete cladding panels too are liable to failure when subjected to fierce heat. High temperatures will build up, causing expansion of the steel reinforcement and the distortion of metal fixings. The concrete will spall and split (thin panels will even buckle), fixings will fail and the cladding will fall readily.

The ability of tongues of flame to reach from one opening to another might be encouraged by combustible cladding materials, but is hardly deterred by the use of noncombustible ones. The greatest deterrent lies in increasing the distance between the openings themselves. Although this would appear to be achieved with a strongly profiled facade (e.g. floors or fins projecting forward of the glazing line), the fact is that such profiles actually turn flames towards the glazing, due to thermal buoyancy creating its own violent pattern of air circulation within partially enclosed spaces.

The beginning of the end for a building on fire might be said to come with the distortion of the main frame by heat. The familiar after-fire picture of the steel frame distorted almost beyond recognition tells a story of extraordinary thermal expansion and forces that can demolish brick walls and concrete floors, even before the fire has reached its peak.

Concrete structures are not immune to such movement. It is possible that the differential thermal movement between steel reinforcement and concrete within a structural member combined with the massive dead loads imposed on it by the structure itself can lead to ultimate collapse. In a very heavy composite structure of steel frame encased in concrete, heat and thermal movement may cause the destruction of the concrete casing but the steel frame will remain to carry the dead load, deferring collapse until it too fails by buckling.

In some ways substantial timber frames may present less risk of building collapse. Although there may be a degree of surface spread of flame over specially treated columns and beams, the charcoal residue of combustion acts as a protective layer to the structural core. Thermal shrinkage in timber will cause less damage to non-combustible wall and floor membranes than thermal expansion in the case of steel. The hazard remains that the surface can transmit flame from one part of the building fabric to another and can transmit fire to spaces that are otherwise separated.

Building spaces

With fuel provided in the building's contents and fabrics, fire needs spaces for the full exercise of its invasive powers. Any spaces in which heat and smoke can be harboured and built up, whether concealed or obvious, small or voluminous, contiguous, communicating or merely leaking, are all significant stepping stones for developing fire.

The presence of an active fire in a building is heralded by advancing smoke. Hot smoke and gases have a thermal buoyancy which, in a confined space, builds up atmospheric pressure above the ambient. For example, in a duct or lobby this pressure will cause smoke and gases to squirt through small apertures, cracks round doors, melted pipe holes and the like, thus transmitting the hazard from one space to the next. In a large space a microclimatic condition occurs, in which a relatively high pressure zone of hot smoke and gas creates air circulation, propelling lethal clouds vertically and horizontally into the uncontaminated lower pressure zones. Hot advancing smoke does not drift – it races.

Any building is essentially an assembly of spaces which can be identified as having either a horizontal or vertical emphasis.

A vertical space, such as a staircase, a lift well or an undivided duct, might be regarded as a flue gathering unburned combustion products and heat, transmitting fire from one level to another by means of convection – the natural upward movement caused by thermal buoyancy. The flue effect can be destroyed in vertical ducts by a separation at each floor. In lifts and staircases this is impractical and it is more important to exhaust the pressure and smoke that otherwise might gather in them.

Bearing in mind how smoke can be propelled by the differential pressures that develop, a horizontal space such as a corridor or ceiling void can also be regarded as a flue. However, because it is more of a trap of heat and volatile gases, it becomes a flue in which flash-over or smoke explosion can occur. As in the case of vertical flues, subdivision by fire-resisting doors or compartmentation can destroy the flue effect. Nevertheless, people have to move through such spaces and the subdivision might not be very efficient and effective. The subdivided spaces are certainly useful in providing safety zones on escape routes, but at the same time they can be impossible to exhaust should they become filled with smoke, except by mechanical means. Thus fire tends to spread at a dramatic pace through horizontal spaces – this is the reason why, in multi-storey buildings, fire will usually spread through the whole of one floor before moving to the next.

The most dangerous of the horizontal spaces to be found in any building are the voids above suspended ceilings. The reasons are many and vital:

- they are the natural trap for hot combustion products;
- they usually contain electrical and other services that may be a source of fire;
- they are frequently formed of combustible materials and contain combustible insulation;
- if insulated, they encourage the rapid build-up of heat;
- they link the extremities of large spaces and may even link adjoining spaces via service outlets;

- they present ideal conditions for flash-over to occur;
- they are concealed – what goes on is out of sight until the ceiling collapses.

The worst aspect of fire in a suspended ceiling is the speed with which it develops and the terrible harm caused by the ceiling collapse. Unfortunately, suspended ceilings are frequently added inexpertly by owners as an afterthought, usually with no knowledge of the potential fire risk. Designers, too, are frequently tempted to specify inexpensive but dangerous materials in areas where risks should never be taken.

In very large spaces, stratification of smoke and gas occurs as hot combustion products rise to the ceiling and expand sideways, cooling as they go until they reach the furthest walls where they fall and are drawn back towards the fire as it sucks in air. This results in layers of black and poisonous smoke moving in opposite directions and at different temperatures. The smoke travels both horizontally and vertically through ducts, ceilings and floor voids, emerging where it can, quite often remote from the seat of the fire. On several occasions in large shopping malls, people have been confused by the movement of smoke and, in an effort to escape, have fled towards the base of the fire. It must not be thought that, because such spaces are large and the smoke trapped within them has the opportunity to cool, they are safe from the danger of smoke explosion. This is far from the truth. Large spaces tend to contain large quantities of potential fuel, which might burn even in a vitiated atmosphere, producing tremendous heat energy. Recently, when a fire in a shopping centre spread to a model shop, the burning of the polystyrene packing round the model kits discharged hot gases into the smoke-logged mall and a peak heat recording of 26 megawatts was recorded. The sprinkler system was capable of containing a fire with no more than 5 megawatts output. Smoke explosion in such circumstances would blast heat and flame into every corner, turning a terrifying smoke trap into an inferno. Enlarge the scale of such an event to that of a massive factory or warehouse in which dangerous products are stored and the result could be a blazing holocaust that might affect a whole community.

Risk and the law

By an elementary understanding of fire, the risk generated by design, construction, use and management of buildings leads to classification and insurance assessment. Consideration of this risk begins with the construction professional in the first basic designs of a building.

However, from time to time the courts have had to consider what is meant in law by fire. This has arisen from disputes over whether damage that has occurred falls within the cover of a fire insurance policy. The usual approach is that ignition must have occurred and that damage caused by heat alone is insignificant.

In *Austin* v. *Drewe* (1816), it was decided that no loss by fire had been caused in the circumstances where sugar, which was insured against fire damage, was damaged by excessive heat from a flue in a refinery where it was stored. Sparks and smoke occurred, but there was no ignition of the sugar.

In *Everett* v. *London Assurance Co.* (1865), the court, in defining the words 'loss or damage associated by fire' in a fire policy, held them to mean 'loss or damage either by ignition of the article consumed, or by ignition of part of the premises where the article is'.

Whether the courts might apply a wider definition, in respect of modern building materials that do not necessarily react to intense heat in the same way as traditional materials, raises interesting questions that cannot be answered with absolute certainty in the absence of direct authority. In any event, 'there is no smoke without fire' is not the attitude taken by the courts.

If the facts of *Austin* v. *Drewe* had been reversed and there had been actual ignition of the sugar, which had caused damage by heat to the building without the building igniting, it would have been held that the building had been damaged by fire. So damage caused to a building by the burning of its contents will be damage by fire even if there is no actual ignition of the building itself.

Placing insurance cover and the duties of the insurance broker are dealt with in later chapters, but in understanding fire the construction professionals should know the legal meaning of the word 'fire', not only in advising the client to ensure that the insurance policy is as broad based as possible to cover all eventualities, but to ensure that a claim is truly covered by the policy wording and is formulated accordingly. In the event of any uncertainty the point should be clarified with underwriters.

'Special plant'

Chapter 3

Assembling the Facts and Counting the Cost

The designer's responsibility

The statutory responsibility for complying with fire prevention rules and providing and maintaining means of escape rests with the owners and occupiers of buildings. However, the designer must never lose sight of the fact that he bears a heavy responsibility over these matters. The brief from the client will, or should, contain comprehensive instructions as to the client's requirements. Although this may touch upon matters of safety, it is unlikely to deal in any depth with fire prevention, fire fighting and means of escape. These aspects of the building must be identified, researched and brought into the design on the initiative of the designer. A designer who fails to take that initiative might well be guilty of negligence.

This initiative regarding fire regulations must be taken while the brief is being compiled. Not only will compliance with the fire regulations influence the design and layout of the building, but it will also have an impact on the cost – not just on the capital cost, which will probably be foremost in the client's mind at this stage, but also on the subsequent cost of the building in use, particularly insurance.

Assembling the facts

The designer will normally be expected to have a reasonable knowledge of the provisions to be made for means of escape and protection, and to have to hand all the relevant Acts, bylaws and

Codes of Practice, all of which can have a profound effect upon the location, design and cost of the building. The designer must therefore gather together all the relevant legislation and regulations, and relate these to the building that is about to be designed. Failure to do this correctly at the start will result in much abortive design work, and time and money will be wasted.

So long as the specific use to which the building is to be put is known to the designer, the full requirements of the relevant legislation can and must be applied. However, the ultimate use or user may not be known, as for example in the case of large multi-purpose building complexes for public assembly or, more commonly, in the case of speculative commercial and industrial developments. In these cases only general requirements can be met in advance, additional provisions being made at the time of occupation or by subsequent alteration. In such situations the designer must be made aware of the need to allow for flexibility in the design of the building, with opportunities to extend circulation or make provision for ventilation and special plant.

At the outset of the design process the designer must not only be acquainted with the relevant Acts, regulations and Codes of Practice, but should also consult on general principles with the fire prevention officer who will be responsible for inspecting and certifying the building on completion. More detailed points of the construction and other matters covered by Building Regulations should be taken up with the local authority's building control officer and, if necessary, with the local licensing officer.

In the case of buildings where there is a possibility that special risks may be involved or might arise in the future, there should be early contact with the insurance company of the owner or future occupier. This will enable the insurer's representatives to identify the risks, to offer advice and also to indicate at an early stage any special requirements the insurers may have.

Insurance and the Loss Prevention Council

The whole subject of fire insurance, the rules that control it and aspects of design and construction affected by it are covered fully in

Chapter 5. Nevertheless, preliminary reference is made here to the subject, which is one of the most important to be considered at the design stage.

Most buildings will be insured by the owners against loss or damage by fire. The sum insured and the terms of the insurance will be agreed by the owner and the insurance company. The insurer will require that the buildings comply with the terms of the Building Regulations and other relevant Acts. However, in a great number of cases these are not specific enough to provide good reason or grounds for the proper insurance. For example, the risk of fire and the magnitude of possible loss is far greater in a flour mill or warehouse with combustible packings than in a private veterinary surgery or community library, although they are subject to control under the same sets of Building Regulations and general legislation. Therefore, to differentiate between the various classes of building and the inherent risks, the insurance companies look to the additional recommendations of the Loss Prevention Council.

Most of the insurance companies in the UK have their own rules to provide the basis for calculating insurance premium rates in all types of building. Although unofficial, they are a supplement to the Building Regulations in those cases in which interpretation or content of the Regulations fail to resolve the needs for safety. The recommendations by the Loss Prevention Council are based on the old rules of the Fire Offices' Committee (FOC Rules), which no longer exist as explained in Chapter 5.

Balancing the cost

At first glance it might seem that very little control can be exercised over the cost of meeting the requirements of the many fire safety regulations applicable to a particular building project. This is far from true, although it must be said that success lies in making the first decision the correct one. For example, an experienced London developer might be heard talking of a site and proposed development requiring a 'Section 20 building'. In fact the kind of building is being identified as one subject to the onerous conditions of Section 20 of the London Buildings Acts, which is concerned with the special provisions to be made in commercial buildings in London over a certain height or volume. The cost of such provisions may seem high in normal circumstances. Seen in the light of prevailing land values, overall development costs and final profitability, they will perhaps appear as an acceptable part of the whole investment, or at least a reasonable balance of cost against risk. With one eye on the regulations and fire safety, a decision on whether to build is the first consideration.

Fundamental design decisions then follow. It is a general rule that the larger the compartment size, the greater is the hazard, the higher will be the standard of fire resistance required both for the compartment and the building structure, the greater the cost and the more onerous are the insurance premiums. Decisions as to the building form may hinge on the most favourable possibility for compartmentation. For the most part the maximum size of compartment is fixed by regulations. In many situations, such as shopping centres where large spaces are necessary, the maximum size of compartment can be doubled provided that sprinklers are installed. This clearly has financial implications.

So it is that many of the principles of fire safety offer the designer a range of choices that will have a bearing on the economics of the building. Some of the broad areas in which cost-effective decisions will be made are listed below:

- *Use of the site* – the manner in which the permitted density or development is achieved; the building's proximity to the site boundary and the impact upon compartmentation, window openings and daylighting; and the consequent availability of useful or lettable floor areas and the resultant return of investment.
- *Building height* – this affects compartmentation, the fire rating of both structure and enclosure, escape routes, lifts and fire brigade access – all factors constituting major constraints on economic viability.
- *Multiple or separate tenancies* – the economic use of floor space, the need to produce lettable floor areas, provide party walls, shared or separate entrance, and the number of staircases.
- *The number and width of escape routes* – this is related to building use; it affects the maximum number of people who may use the

building (or floor), the subdivision of floors, travel distance, dead ends and, as a consequence, the whole floor shape.

- *Choice of structure* – choices include heavy or light weight, wet or dry, traditional and economic or high speed and costly with early return on investment; fire rating is a factor.
- *Design of elevations and fire rating of external materials* – the surface spread of flame, location and size of window openings, sill heights, degree of natural ventilation and thermal performance, all of which are economic matters relevant to speed of erection.
- *Roof construction* – form and choice of materials, their resistance to fire and spread of flame, proximity to adjoining hazards, access and maintenance, thermal performance and condensation.
- *Surveillance, detection and alarm systems* – with economic advantage these can affect the provision for means of escape and separation of fire risk areas.
- *Automatic sprinklers and other prevention systems* – these are costly provisions that may be necessary to the use of the building but might otherwise allow economy in compartmentation and greater flexibility in the use of space.
- *Smoke exhaust systems* – these may be necessary for certain public buildings but otherwise can affect the building form, the number and design of escape routes and staircases, all with overall economic advantage.

Refurbishment

The most complex tasks involved in refurbishing a building will be to bring compliance with fire regulations and means of escape up to date and to house the plant, machinery and electrical installation safely and out of sight, all to modern standards. It is this aspect, rather than the cost of repairing or modernising the shell, which is usually responsible for making the total cost of refurbishment almost as much, sometimes more, than the cost of an equivalent new building.

However, other factors usually come into play in deciding whether to rebuild. In the case of historic or listed buildings, or in meeting conservation area requirements, planning considerations take precedence over pure economics.

With historic buildings there is a body of experience continually devising ways to improve standards of fire resistance without destroying important features. Not all of them are successful – in general a compromise to a reasonable standard is achieved, as an improvement on previous conditions. Parts that are totally rebuilt can, by careful design, comply with modern standards. Where common sense prevails, building control officers can help by suggesting the relaxation of the regulations' requirements. Early consultation is a key to success in this respect.

Priority of facts relevant to legislation and insurance

In referring to legislation or conferring with design specialists, it is helpful to have first prioritised all those facts about the building which may seem relevant to fire protection, statutory control and insurance. It will follow that as work or consultation proceeds many points will emerge as additions to the brief, influencing the design as it is developed. It may not be possible to anticipate every facet of a building that might have such relevance, particularly as far as insurance is concerned.

The list of legislation is daunting and far beyond the province of this book, but becomes less so once the principles of defence against fire are understood in broad terms. Even then, the extensive legislation can be looked upon as a source of guidance rather than restriction. Counting the cost of building safety and economy is a matter of experience, bearing in mind that:

- legislation can mean choice and may therefore allow cost analysis and comparison;
- insurers can contribute valuable information at an early stage that might effect substantial savings in premiums during the life of the building;
- economic safety can be achieved by understanding how fire develops and by good design, without necessarily implying greater initial cost.

Immunity of liability for local authorities

In the last third of the twentieth century, the law has attempted on numerous occasions to identify responsibility of local authorities in the approval of construction design and the inspection of construction on site. No attempt is made in this book to list the cases concerned with this or to comment on the decisions of the courts. Suffice it to note that, while in *Anns* v. *Merton London Borough Council* (1978) the House of Lords held that a local authority owed a duty of care to avoid causing a present and imminent danger to health and safety in the discharge of its obligations under the Public Health Acts, in 1990 in *Murphy* v. *Brentwood District Council* [1991] the House of Lords departed from *Anns* in finding that a local authority owes a duty of care to occupiers of properties only to ensure that they or their property (other than the building itself) do not suffer actual physical injury.

The situation of the responsibility by the local authority owing a common law duty of care to avoid loss of or damage to a property by fire is illustrated in the case of *Tesco Stores Ltd* v. *Wards Construction (Investment) Ltd* (1995). The brief facts of the incident revolved around a small shopping complex constructed in 1987–8, comprising shop units, public house, supermarket and ancillary facilities. In 1993, vandals caused a fire to start in a rubbish bin outside one of the shop units. The fire spread into the shop unit and then rapidly into the supermarket, and it took 25 fire appliances some six hours to bring the fire under control. Severe damage occurred to the buildings and the stock but there was no loss of life or injury.

The details of the ensuing legal case are complicated in that numerous other parties were involved, including the Maidstone Borough Council. A preliminary issue was ordered to determine whether on certain assumed facts the council owed a common law duty of care to avoid causing the plaintiffs loss of or damage to property. Against the council it was alleged that a duty of care was owed by the council to the plaintiffs (who were tenants of the building) in its approval of design drawings and inspection of construction to ensure that proper fire prevention precautions were taken and that the relevant Building Regulations were complied with, and that this duty had been breached.

The case for the plaintiffs failed and in essence the court held that the Building Regulations (1985) did not require anything to be done beyond what is necessary to secure reasonable standards of health and safety for persons. It further considered that it was appropriate to distinguish between regulations the purpose of which is to prevent damage to health, safety or welfare of persons and regulations the purpose of which is to prevent damage to property.

The judgment confirmed that the Building Regulations were concerned with the health, safety and welfare of persons, and it was not their task to protect property or chattels. It therefore followed that in the circumstances of this case, a local authority does not owe a common law duty of care to avoid causing to owners or occupiers loss of or damage to property. It concluded that, even if the purpose of statutory requirements was to avoid damage to property, on the assumed facts of this case it was not fair, just and reasonable to impose a duty of care at common law on local authorities.

Personal injury is not part of this case nor part of this book. Responsibility or liability for material damage from fire, however, appears not to rest with local authorities. The construction team from the designer to the builder must ensure, when assembling the facts and counting the cost, that owners, tenants and insurers for the property have no action against them for failing to take reasonable precautions in the design and construction of the property to prevent fire. This would include complying with the requirements of the current Building Regulations and any further measures related to the specific use, situation and construction of the building.

'. . . the obvious place in which to look for fire-fighting equipment must turn out to be the place where it is actually to be found . . .'

Chapter 4
The Means of Defence

Design

Having understood the nature of fire, cause and effect, and assembled the facts and costs particular to the project, the first risk is presented: that of providing within the design of the building the very best means of defence against loss of life and material damage from fire. In addition, the designer must assume the design obligations arising from the Construction, Design and Management (CDM) Regulations; failure to do so may be very expensive.

Fire threatens both life and property and covers a wide range of losses. Among these are the capital value of material damage by fire, loss of employment in industrial and commercial properties and environmental damage caused by fire and fire fighting activities. Most laws and regulations are directed to protect life; financial protection of property against fire loss is a matter for insurance, but the designer can greatly influence that risk.

The maintenance of safety and protection against loss – the purpose of defence – calls for a strategy in design that recognises:

- the nature of fire
- the performance and structure of materials
- the behaviour of people.

The strategy is founded on the principle of instant understanding of the building by those who use it. The designer must think 'fire' at all times.

- The design must be clear.
- The fabric of the building must be the safest possible.
- The apparent ways in and out must indeed be the correct ones.
- The natural direction in which to escape must turn out to be the safe one.
- The obvious place in which to look for fire fighting equipment must turn out to be the place where it is actually to be found.
- The equipment when found must be right for the job and in working order.
- The fireman's natural approach to a fire must turn out to be the best one.

The main defensive measures fall into two main categories: passive and active. Passive measures are inherently safe, in-built characteristics, which are effective by their presence. These include:

- clarity of design
- good access
- simple circulation
- good means of escape
- compartmentation
- doors
- protection of the structure
- resistance of materials against spread of flame
- ventilation.

Active measures are those that come into use when fire breaks out:

- detection and alarm systems
- sprinkler systems
- fusible link doors and shutters
- emergency lighting
- smoke exhaust.

Passive measures

Once again, it has to be stated that legislation, regulations and all documentation related to fire appear to be endless and under day-to-day revision. Basic principles of design, however, remain fairly constant. The following lists are illustrative, but not comprehensive.

The building on its site

- Use and planning requirements;
- relationship with its neighbours;
- the distance between a building and its site boundary or an adjoining building;
- the manner in which the building volume is arranged;
- undefended spaces between separate buildings where doors and windows facing each other, which can create fire passages;
- narrow spaces between or within the building, which may behave as a flue;
- clarity of design in siting and massing the building at the initial stage.

The building form

- Purpose;
- size;
- separation or division;
- resistance to fire of the selection of materials in the structure.

Good access

- Adequate provision of access for the fire brigade to reach all elevations;
- room to manoeuvre appliances;
- sufficient depth of space from the face of the building to erect ladders;
- restriction of parked vehicles on the pre-determined fire paths for fire appliance access;
- unobstructed access for fire appliances of all types including turntable ladders or hydraulic platforms, and providing gentle gradients free of bollards, lamp standards, trees, signs and other obstructions.

Circulation and means of escape

- Clear vision of the direction of movement within the building;
- easy sight of obviously located stairs and lifts;
- a clear sense of location at all times by the users of the building;
- a natural response and awareness to the position of entrances and exits to the building;
- width of escape routes related to the use of the building and time allowed for evacuation;
- consideration of the number of ground floor external exit doors, in order to allow simultaneous discharge of the total occupants in the prescribed time, while catering for crowd panic and failure of some doors to open or function properly;
- consideration of external space to accommodate the evacuation, and defined external areas for assembly to ensure accountability of all occupants.

Compartmentation

- Limitation of building volume related to its use, to comply with the regulations;
- fire rating of structural elements;
- use of fusible link doors and shutters to provide communication between compartments;
- protection of perforations to accommodate services;
- intumescent strips to openings and louvres;
- fusible link dampers to ventilation trunking;
- limitation of building volume to meet costs of construction and lower insurance premiums, compared with restricted use of the building;
- the smaller the volume, the less the fire load.

Doors

- The function of doors as a barrier to fire and smoke;
- design of doors to include their frames, glazing and ironmongery as one combined element as 'door sets';

- door sets to be considered at design in the context of their use and location:
  - in regular use but closed between times
  - in constant use, which need to be kept open until a fire occurs
  - only used for very special occasions such as access or maintenance to ducts
  - used for escape only, which might otherwise present a security problem
  - located on escape routes and having the dual purpose of smoke containment and fire protection
  - when closed form part of the compartment wall or enclosure, with which they need to provide an equal fire resistance;
- glazing to doors to provide clear vision for fire fighters to see inside rooms, yet to be able to resist the movement of heat and smoke so long as the frame holds up.

Protection of the structure

- Fire rating of each material to be used;
- thickness of protection cover to structural members;
- fire resistance of each compound structural element;
- fire retardant solutions;
- method of assembly and integrity of fixings;
- stability of structural elements through fire and smoke;
- resistance of each material to spread of flame;
- restrictions to spread of volatile gases from burning contents;
- restriction to spread of fire and smoke in hidden voids in roofs, suspended ceilings, raised access floors, ducting and similar passages;
- selection of elements of the building to withstand the risk of ignition from falling burning debris.

General precautions applying to passive measures

The passive measures to defeat the spread of fire apply to almost every element in the building. Before final selection, the choice of each material to be specified should be reviewed in terms of its resistance to the spread of flame. Similarly, hidden spaces beneath floors, above ceilings, in service ducts and in rooms should be carefully considered, so that barriers, protection and fire stops are provided and, where possible, non-combustible materials are used in order to resist fire, smoke and the spread of flame.

Ventilation is necessary in most buildings and this can be provided in a number of ways. Both passive and active measures may be used, together with clear and unambiguous signage, to guard against or to reduce the risk arising from the spread of fire. Preservation of fire integrity should be considered at each design stage to ensure that the choice and specification will reduce the opportunity for fire to occur and, should it happen, will prevent it from spreading and allow easy escape.

Active measures

When a fire occurs there ought to follow a chain of events that would justify the description of 'the law of active measures'. Hence:

Fire + Detection = Alarm + Protection.

Alarm

- Local alarm: visual and audio;
- remote alarm: fire brigade and police.

Protection

- Electrical control: standby power, lighting and lifts;
- automatic fire control: sprinklers, water spray, deluge, gas and powder systems;
- automatic smoke control: ventilating and pressure systems;
- automatic isolation: doors and shutters;
- local fire fighting: dry and wet risers and hand appliances;

- public services: fire brigade, ambulance services and police control.

From the moment of fire detection, many of the activities listed occur simultaneously. Alternatively, with certain hazards or particular conditions of escape, reaction to fire detection may be programmed to occur in a particular sequence, be phased, or have in-built delays.

Detection and alarm

Depending on the building's use, size, value of contents or number of occupants, detection may be represented by simple local response equipment, as in domestic premises, or by a total surveillance system designed to meet the particular needs of a large building or complex such as a major hotel, factory or warehousing facility. Such systems might incorporate a wide range of detection equipment, each type selected for its suitability to operate in the local conditions and each one signalling to a central control or monitoring point, from which the appropriate warning and protective measures will be activated. Signalling can be extended via telephone lines to an off-the-premises monitoring station run under a full-time surveillance contract or, in special circumstances, direct to the fire brigade.

A variety of detection mechanisms are used which involve the completion of an electric circuit as a result of the thermal expansion of metal, liquid or gas, or as a result of the break in a fusible link. Other types involve the change in the electrical resistance of conductors, or the change in voltage produced between two thermocouples.

Heat detectors

- Fixed temperature – designed to operate at a pre-selected temperature;
- rate of rise/fixed temperature detectors – operate when an abnormal rise of temperature occurs or when a pre-selected level is reached.

Smoke detectors

- *Ionisation detector.* A chamber containing electrodes, connected to a supply. A small radioactive source ionises the air in the chamber, turning it into electrically charged particles (ions) and so allowing an electric current to flow. Smoke particles attach to some of the ions and so reduce the current, activating the alarm.
- *Optical or thermo-electric detectors.* A light beam passes close to or falls on a photo-electric cell. Smoke entering the detector scatters the beam, changing the amount of light that falls on the cell and setting off the alarm.

Combined heat and smoke detectors

- *Infra-red detector.* This consists of an emitter transmitting a wide angled beam of infra-red light that falls on a remote photo sensor. Loss of strength in the beam caused by smoke or thermal turbulence registers a fire signal upon exceeding a pre-determined level. Correction for temperature or accidental blockage can be built in.
- *Laser beam detector.* This consists of a projector sending out a narrow laser beam that is reflected on to a photo-cell receiver, partially masked so as to contain the incident spot of light in a 'no fire' position. The alarm is activated when the beam is deflected on and off the spot by the slightest turbulence from heat or smoke particles, but it responds only to the frequencies which would be given by a fire.

Flame detectors

- *Infra-red, scanning or defined area detector.* This is a photo-electric cell, the frequency of which is fixed to respond to the low-flicker infra-red radiation characteristic of a diffusion flame. Scanning detectors rest in position when a signal is picked up. If the radiation flicker pattern persists, then the alarm is given; if not, the scanner moves on.
- *Ultra-violet detector.* As for the infra-red detector, this detects the ultra-violet light emitted from flames.

The choice of detector will depend upon circumstances and, very often, a number of different types will be required in combination to deal with the varying conditions found in one building or even within a single space.

Alarm

In large automatic systems, detectors will be grouped in zones so that the location of the fire can be identified. For the use of personnel, in or escaping from the area of an otherwise undetected fire, it is essential that manually operated alarms are also located in the same zones and linked with the automatic detectors to the indicator panel.

The alarm itself might signal immediate total evacuation, as in a place of public assembly, or it might be phased to give a general standby warning (indicated by an intermittent call sign) together with an immediate evacuation instruction (a continuous call sign) from the area of first danger. Such systems minimise the rush to escape routes and give time for the proper closing down of machinery and organising the grouping of personnel at their allotted fire stations. Indeed, in some establishments it would be ill-advised for occupants to leave their workplace unless danger of fire reached a degree of imminence. Where the system is linked to a remote monitoring station, or direct to the fire brigade, the alarm would be activated at that location and, again, the area of the fire would be identified.

The alarm signal must be one that can be readily identified by occupants, especially against a background of other intrusive noise such as machinery. Visual alarm signals should show concurrently with audio alarms, particularly where working conditions call for the use of ear muffs. The deaf must not be forgotten.

Protection

The active measures to provide protection and to reduce the risk of injury to person and material damage from fire should embrace the following.

Electrical supply

There is a moment's pause between the cut-off of normal supplies and cut-in of standby generators. This temporary cessation in the supply will affect other plant and machinery, temporarily stopping lifts, air conditioning and exhaust systems, cutting out electric lighting, releasing magnetically restrained fire doors and so on; all likely to cause confusion and anxiety. It is important that proper phasing is built into the electrical control with manual override switch gear for operation by appointed staff or by the fire brigade. The manual override must be located in an obvious position and be easily accessible. It must be possible for parts or the whole of the building to be isolated from all electrical supply if there is any risk of this contributing to the danger. It is possible that the cause of the fire may be electrical or that fire might cause fusing, overloading and overheating, or sparking in the presence of explosive gases, hence adding to an existing fire.

Escape and emergency lighting

Emergency lighting is required to provide a minimum level of illumination in the event of failure of the normal lighting. Escape lighting is the part of an emergency system that ensures that the means of escape can be identified and used effectively. Along escape routes, the frequency of lighting must be sufficient to achieve the minimum level of illumination between each fitting, with special emphasis lighting provided at danger points such as bends in corridors, doors and staircases. Luminaires may be used as guides to follow the escape route and ideally each luminaire should in some way indicate the position of the next. Exit and emergency exit signs should be positioned immediately above or adjacent to the exit door. Where this is not in direct line of sight, subsidiary signs should be used to lead people in the right direction.

Lifts

Lifts should be automatically isolated from remote control once a fire alarm has been given. Lifts can continue to be operated by trained fire fighters and in a large building with banks of lifts at least two will need to have fire service control. The fireman's switch, located in a position recommended by the fire brigade, will only be operated by an authorised person who will use the lift from within the car. All lifts should have an emergency telephone and emergency lighting.

Sprinkler systems

An automatic sprinkler system is not merely a means of pouring water on a fire. It combines detection, warning and restriction. Essentially, it is a first-aid treatment, spotting the fire, spraying water on and around it, cooling the surrounding fabric and possibly extinguishing the blaze. At the same time it will sound the alarm and indicate the zone in which the fire has occurred. The exact type of system required will depend upon:

- the nature of the contents being protected and whether or not they might be seriously harmed by a leak or accidental operation;
- the degree of hazard generated by the contents and the rate of water discharge required to make the fire area safe.

There are many systems of sprinkler installation, each of which will need to comply with the recommendations of the Loss Prevention Council (LPC) for automatic sprinkler installation. It is advised that installation and maintenance should be executed by an LPC Certificated Sprinkler Installer. Among the systems which are in current use are:

- traditional wet
- alternate
- pre-action
- cycling
- on-off
- deluge
- high velocity spray.

Innumerable variations on these systems can be introduced to deal with the special requirements of a building or its contents. Even though the design of sprinklers is a highly specialised matter, knowledge and forethought on the designer's part will ensure that the general or special conditions for their installation are catered for at an early stage in the design process. To this extent a check list of points to remember is useful.

- The adequacy of the water supply must be checked.
- Standby generators or batteries may be required to maintain the operation of controls and pumps.
- In any one building, more than one type of system may be required, e.g. offices in special warehouses.
- Sprinklers will be required:
  - over escape doors;
  - over escape and other stairs;
  - in lift shafts;
  - in stores and some plant rooms;
  - possibly in ceiling spaces and ducts.
- Sprinklers should not be installed in electric intake rooms.
- Sprinklers may be required at several levels inside high storage racking or within special machinery.
- To avoid unnecessary damage, the water from sprinklers will need to be disposed of either with scuppers to the exterior or by a system of floor drains.
- Points for drain-off at low level and pressure testing at high level will be required.
- In large installations the main distribution pipes may be of surprising size and require careful co-ordination with other services.
- Thermal or structural movement joints may be required in main pipe runs.
- Access to all parts for regular inspection and maintenance is a necessity.

A final warning to designers: the operation of sprinklers can cool a developed fire to a state of partial combustion so that heat-operated ventilation systems fail to come into action and dangerous smoke logging occurs. Conversely, the prior operation of an automatic smoke exhaust system can increase the circulation of air and intensify the heat of a fire, making it more difficult for the sprinklers to control the situation.

Foam injection

The use of foams is generally associated with heavy mobile fire fighting equipment or small hand appliances. In fact, foaming agent injection equipment can be fitted to water systems such as sprinklers or, more commonly, wet riser branch pipes and hand hose outlets. Without doubt, foam is the most rapid and effective medium for fighting fires involving flammable liquids and toxic chemicals – so rapid is the effect of complete saturation in foam that the expression 'knock down' is used to describe the extinguishing of a fire. The success of foam depends upon its low viscosity and stability, enabling it to form an insulating blanket that will smother fire, allow time to cool, prevent 'burn back' (i.e. re-ignition) and suppress the release of noxious combustion products. Foam can be stored in large mobile fire fighting appliances which are kept in a state of readiness close to a likely source of fire, such as airport facilities, large fuel and chemical storage plant or fuel handling equipment. Built-in foaming plant can be installed for the permanent protection of stationary hazards such as floating top fuel tanks and fuelling bays, these requiring regular and careful servicing. On a smaller scale, hand appliances have application in cooking areas, garages and certain types of plant rooms.

Gas and powder systems

Actuated by the same methods as sprinkler systems (smoke and heat detectors), built-in gas and powder systems also combine alarm and defence. They are used in situations where water spray or deluge systems might not be effective, could be dangerous or would cause damage to equipment and materials. Typical of such circumstances are processes or plant involving water reactive chemicals, spaces for delicate electronic equipment (computer rooms) and machinery spaces for the manufacture of delicate products. Even within the range of gas and powder systems, however, the use of one or the other could prove harmful. For example, carbon dioxide can cause freezing up of metallic parts in delicate equipment; powder can clog moving parts, conceal damage and be difficult to clean.

Gaseous and powder agents are of two types.

- *Inerting agents*. These suppress fire by the physical means of displacing the air in the area of combustion, cutting off the oxygen supply, and by reducing the temperature to below the combustion level. The use of carbon dioxide is most effective against hydrocarbon fuel and electric fires, in enclosed spaces that may be safely filled with gas, or in small concentrations in restricted work spaces. Typical applications will include:
  - engine room spaces involving petrol or oil fuel;
  - oil-filled switch gear, transformers, alternators, in power stations and substations;
  - computer suites and control rooms housing electronic equipment;
  - accommodation and storage for valuable or delicate goods or works of art, which would be vulnerable to water.
- *Inhibiting agents*. These combine inertion with chemical suppression, so requiring a less concentrated application than inerting agents. There is a wide range of halogenated hydrocarbon agents, or halons as they are known, all with confusingly similar and lengthy names. Halon is a dry, colourless, odourless, noncorrosive and nonconductive gas; it is an inhibiting agent that separates the elements of fire, causing a breakdown of the combustion process. There are certain rules to be observed in the use of halons:
  - All systems must be custom designed to suit the individual application (hazard, location and occupancy).
  - Flooding systems are designed to operate in enclosed spaces and all openings, air extract plant, etc., must be closed prior to release of the agent. Spaces must be sealed to prevent escape of the extinguishing agent or combustion products and subsequently well ventilated before re-entry.
  - Halon gas is odourless (unless scented), so that release by an automatic fire alarm system must only be possible when an area is unoccupied or where a two-stage warning system is installed.
  - Automatic release systems must have manual override control with status indicators on a central control panel. Where local application of halon is required, storage will be in small wall- or floor-mounted cylinders with the minimum pipe run to a single nozzle outlet. Examples of this type of application include paint spray booths, engine test cells, laboratory

cabinets, quench tanks, dip tanks and oil-filled transformers. Flooding applications are used where a hazard is within a fixed enclosure, which will ensure that an adequate extinguishing concentration of agent can be achieved and maintained for an appropriate period of time; for example, cable voids, motor rooms, relay rooms, computer rooms, tape and record stores, art galleries and libraries.

Powder systems (inerting)

Fire fighting powders are finely ground chemicals, constituents of which depend upon the type of fire to be extinguished. Powder may be used in hand appliances and it is vital to note before use whether they are suitable for extinguishing or controlling the classification of the fire:

- class A – solid fuel fires
- class B – flammable liquids
- class C – combustible gas fires
- class D – flammable metal fires.

Powders should not be used on:

- low-voltage electrical installations, relays, switch gear, generators or motor equipment;
- any form of telecommunications equipment, radio, TV or similar installations;
- computers, microelectronic equipment and associated hardware and software storage;
- tightly packed solid combustibles where deep seated fires might occur, such as record files, furniture, fabrics, furs, etc.

While this appears to be a formidable list of areas in which powder may not be used, those in which it is effective and suitable are far more numerous and only a few need to be mentioned:

- boilerhouses and engine rooms;
- oil storage tanks and oil rig modules;

- domestic and commercial kitchen ranges, hoods, ducts and general service areas;
- industrial installations involving milling, pneumatic conveyors, dip tanks and quenching baths.

Mechanically assisted ventilation systems

Apart from permanent natural ventilation there are various mechanical options open to the designer. Selection depends upon the problems created by the building design and the degree of risk. All such systems can be activated electrically by remote heat or smoke detectors or from a fire control panel, or they may be triggered when heat breaks a fusible link.

- *Mechanically opened natural vents* (invariably roof mounted). These are opened by the lever action of counterweights or by pneumatic or hydraulic action.
- *Smoke extracting fans*. These are usually roof or high-level wall mounted and in some circumstances associated with ducts of limited length. Fans can be used for natural ventilation and cooling but have an automatic cut-in activated by the fire detection system. Extract fans are the simplest means of preventing smoke logging of large spaces where there is direct contact with the exterior. The air velocity must be sufficient to overcome external wind pressures, but the air circulation so created can stimulate fire.
- *Ducting extract systems*. These are most likely to be part of a building's mechanical ventilation system. While of great value in evacuating smoke from special areas and spaces deep inside a building, danger lies in the possibility of spreading smoke via the duct system to unaffected parts of the building or of a fire in the ducts themselves. Air pressures at discharge points are unlikely to be sufficient to overcome adverse wind conditions. Good maintenance, always difficult, is a prerequisite.

Pressurisation systems

Designers of large buildings will undoubtedly become more involved in this form of smoke control and protection, but it must be remembered that this is a field of work requiring the services of specialist consultants. However, guidelines for the designer to bear in mind are listed below.

- Highest air pressure is at the point of greatest safety; the lowest pressure at the point from which the first steps towards escape must be taken.
- Lift shafts, lobbies, ducts and lavatories on escape routes can cause diversion of air flow. Air input and pressure must be adjusted accordingly.
- The fans supplying air under pressure must be duplicated with standby units, all with automatic control and manual override.
- Air turbulence and contra-flow is created by the opening and closing of doors; external microclimatic conditions can disrupt exhaust. Air pressure and velocity must be capable of overcoming these problems.
- Doors must swing against the direction of air flow (which will also be the direction of escape). Closers will not be necessary, except where pressures are very low.

Automatic control of doors and shutters

Smoke control and containment of fire is normally achieved by arrangements of doors and shutters such that they are held in a closed position when not in use. There are, of course, a great number of circumstances where doors are required to be kept open until the moment a fire occurs, otherwise they may interfere with working operations, sales activities or circulation in general. The range of door and shutter types, and the electrical or mechanical equipment operating them, is wide; almost any combination that might resolve the designer's difficulties can be fabricated.

Control of a fire door's motive or release mechanism will be by one of the following:

- a fusible link responding to local heat, so as to release a counterweight system or interrupt an electrical circuit operating magnetic catches
- a solenoid switch activated by a local heat or smoke detector or by

a general detection and alarm system, so as to set in motion electrically motorised machinery.

All doors that are not in normal use must be capable of manual operation. The following are some of the doors available, giving the manner of their operation and circumstances in which they might be used.

- *Automatic sliding doors (usually glass).* These are beam or mat-switch controlled; they remain closed until approached and therefore remain an effective barrier to smoke. They are a high-quality installation, used in airports, hotels and shops.
- *Hinged doors.* These are held open by a magnetic catch, released by interruption of electric circuits by fusible link or solenoid switch, and closed by standard door head or floor closer. They are used widely in hospitals, offices and old persons' homes.
- *Slide folding steel doors.* Motor-operated types have solenoid switches activated by the alarm or detection system; manually operated types have an emergency counterweight closing device on a fusible link. Doors can be manufactured to four-hour fire rating and are therefore useful in compartment walls. They are used mainly in industrial situations where doors are in constant use and the floor track will be kept clean.
- *Vertical sliding steel doors.* These are motor operated as above, with solenoid switch or gravity action with counterweights and fusible link. These doors are reliable and can be manufactured to four-hour fire rating. They are used in industrial and commercial situations where they will be held open at all normal times (e.g. supermarkets, shopping malls).
- *Horizontal sliding steel doors.* These are very heavy duty and are therefore motor operated with solenoid switch control. Advantage lies in their high potential fire rating but there is a disadvantage in their need for a floor track, which might not always be kept clean. They are used in industrial or commercial situations where there is no headroom for vertical sliding doors and where they will be open for most of the time.
- *Roller shutter doors and escalator shutters.* These are normally motor operated with a solenoid switch control activated by a detection or alarm system. Manual operation is normally by chain purchase

on the roller spindle. Fire rating is limited by size but roller shutters are useful in all situations where there is a shortage of headroom or floor tracks are a disadvantage, and where intruder security is not important.
- *Overhead doors*. These are generally lightweight and used for domestic garage purposes, having little fire protection. In industrial circumstances they would normally be motor operated with emergency cut-in solenoid switching.

All electrically operated doors must be on a standby electrical supply circuit if this is provided. The only doors that will function at the time of complete electrical failure are those restrained by magnetic catches and those that operate on counterweight systems (provided the fusible link is broken). It must be restated that completely automatic systems are not acceptable and that manual operation should be possible. Occupants might be cut off from escape by doors that have closed; a critical factor in the public areas where heavy-duty doors are used. Small personnel doors built into heavy doors of overhead, vertical or horizontal sliding type would allow this danger to be avoided.

Wet and dry risers

Wet and dry risers are installed to bring a supply of water close to the seat of a fire so that it may be tackled from inside the building. They are appropriate to a wide variety of buildings, or particular areas of buildings, where sprinklers or other systems might prove unacceptable or ineffective in protecting life, e.g. schools, hospitals, auditoria, small hotels, offices and other commercial or industrial situations. They are suitable where carbonaceous fires are involved (e.g. wood, paper, textiles and furnishings).

For their simplicity and efficiency wet risers are preferred where the existing water pressure is sufficient for their proper operation. A fully charged water supply pipe has a hose reel permanently connected to it. When the inlet valve on the supply pipe is opened, the hose reel is immediately ready for use; the nozzle is removed from the bracket and the hose run out towards the fire. The nozzle can then be opened to give a 9m jet or a spray. Ideally, the water supply for hose reels should be separate from, and independent of, the general water supply to the building. Wet risers and reels should be positioned so that the hose, when run out, will not impede the means of escape or prevent the proper closing of smoke stop doors. They should not therefore be placed on staircase landings. Siting should be on every floor level, adjacent to the point of exit and very obvious. The reel, which can swing out at an angle to the wall, may be recessed so as not to obstruct the corridor or escape route.

Dry risers are installed mainly in multi-storey commercial or residential buildings, where large quantities of water are required under very high pressure, but where the existing mains pressure is inadequate to reach all levels and where use is to be strictly limited to that of the fire brigade.

A heavy duty uncharged service pipe is provided to all floors from an obviously accessible coupling and valve, ready for connection with a firefighters' hose to a high pressure supply. Once connected, the pipe is charged by opening the valve, the air being cleared from it via a high-pressure valve at its head. Single or smaller double couplings with gate valves may be provided at each floor, to which firefighters' hoses are connected and run off to the location of the fire. Since the hoses have to be taken to the level of the fire it is wise to locate the dry riser close to a lift under firefighters' control.

The virtually unlimited length of hose which might be used means that, in most situations, only one dry riser is necessary in a building for water to be taken to almost any point. Additional dry risers may be provided for more simple strategic reasons, e.g. difficulties of fire brigade access at ground level, extremely extended floor plans or the intervention of compartment walls with automatically closing doors isolating wings of a building. It will be necessary for the designer to consult the fire officer on such matters, beyond which the rules of location are similar to those for wet risers.

Hand-held extinguishers and other equipment

Beside the full gamut of built-in protection systems, the small hand-held fire fighting equipment that should be provided may seem little more than incidental paraphernalia. This is not so. Although it must be insisted that priority goes to the evacuation of the building, widespread fire can be prevented by immediate action on the part of occupants who are on the spot and who can find a suitable extinguisher ready to hand. The types of extinguishers which must be made available for emergency include:

- water extinguishers (water under pressure);
- foam;
- carbon dioxide;
- halon gas;
- powder.

All are contained under pressure in cylinders of assorted size, ranging from total discharge water, foam and powder extinguishers for general purpose (usually hung low on walls because of their heavy weight) to the latest halon mix extinguishers which, like standard aerosols, are of controlled discharge and very compact.

The criteria for the choice of extinguishing agent are exactly as previously described for in-built systems. All containers describe the contents and the appropriate use.

In a slightly different category, mention must also be made of fire blankets. Originally made of asbestos but now of heavily woven glass fibre, these are rolled tightly and contained in open-ended cylinders from which they can be quickly pulled in an emergency. These are used for class A and B fires and are particularly useful in dealing with cooking fires.

In many circumstances the provision of hand equipment is left to the building user, at the direction of the local fire prevention officer, after the building is completed. However, this procedure invariably results in difficulties in finding suitable locations, or locations in which the appliances are not a physical or visual intrusion and consequently later damaged or moved. Proper positions are:

- in staircases and entry lobbies leading to primary circulation routes, so that they can be collected on the way in to attack a fire;
- adjacent to escape doors, so that they can be used on the way out;
- at focal points in work areas and circulation routes – intermediate resources seen from more than one direction;
- adjacent to special hazards, e.g. electrical switch gear, machinery of all descriptions;

- outside stores of dangerous combustibles, e.g. chemicals, liquid fuels, organic materials liable to spontaneous combustion;
- adjacent to special work stations, e.g. in laboratories, many types of manufacturing facilities, cooking ranges.

In key locations fire extinguishers are best sited en suite with other fire fighting and protection equipment, in specially designed and well-displayed fire fighting stations. The outlets for dry and wet risers, together with their hose reels, are available in neatly designed units which, if required in the building, would form the basis for such fire fighting stations. Other items of equipment that might also be included are:

- fire alarm and call point
- emergency telephone
- first-aid supplies
- stretchers
- protective clothing
- breathing apparatus
- portable escape apparatus such as ladders, chutes, etc.

Maintenance of extinguishers and other equipment listed above can be an onerous task in large buildings or complexes. Plans should therefore be available which show the location of the fire fighting stations, with schedules of the appliances giving detailed records of their servicing.

Public services

Although fire, ambulance and police services can be called through public emergency telephone, where the risk is complex or concentrated, management should ensure a detailed inspection by each service and take advice on recommendations for possible future fires and major incidents. A contingency plan should be determined and put in place with practice rehearsals for building evacuation carried out at regular intervals. These contingency plans should provide not only for unassisted evacuation of personnel but for the provision of the injured and trapped. Although many of these plans would be regarded as management responsibilities, the design and provision of the passive and active means of defence must enable management to be given the proper opportunity to function in a highly adequate manner.

'. . . someone to shout FIRE . . .'

Chapter 5
Design and Insurance Requirements

Annual expense of fire insurance

The design of a building should impose on the architect the duty to provide a structure that not only complies with all statutory regulations but will give the user the benefit of economical maintenance and cost in use. Considerable thought is usually given to this subject and much time is spent on balancing capital expenditure with maintenance costs on items such as heating installation, windows, double glazing, self-finish surfaces and the use of many maintenance-free materials. However, little time is devoted to the cost of fire insurance premiums, which recur annually and are ever increasing. Architects may suppose that they have no control over this matter; this is untrue.

Fire insurance premiums on buildings can be as low as 0.1% of the sum at risk. Where the risk of fire is great this premium can increase manifold and poor design can be significant in this assessment.

Statutory requirements and insurance objectives

Most statutory requirements are primarily directed towards safety and preservation of life. They influence the design of the building by reducing the possibility of fire occurring and, in the event of fire, by restricting its spread and providing adequate means of escape. The insurer is concerned mainly with the protection of the property from destruction or damage by fire and the preservation of capital and material assets. One of the important factors determining the premium rate of fire insurance for a building is the design of the structure to restrict damage in the event of fire.

There are other factors that greatly influence the premium rate. The use, structural condition and location of the building will all be taken into account by the insurance surveyor when preparing a general appraisal and assessment of the risk that the company will be underwriting. The history of the insured in respect of past insurance matters will also be a consideration of the underwriter. Whereas matters of pure insurance are outside the scope of this book, the design of the building is the responsibility of the architect and he should consider seriously the fire risks to the material asset in his basic concepts.

The designer's attention must also be drawn to the inadequacy of the Building Regulations regarding fire precautions and consequently the cost of fire insurance premiums. This can best be achieved at design or construction stage of the building. As far as the Building Regulations are concerned, no distinction is made between a warehouse storing polystyrene goods and another storing steel beams; yet the fire risk is entirely different and fire insurance premiums can be extremely high unless the design and construction take this into account.

Insurance premiums are related to risk. In respect of buildings, the fire risk relates to the material asset. Premiums for fire insurance of buildings are the charges for the considered business risk undertaken by the insurers relating to the value of the material assets at stake. Risk improvement, therefore, is a positive way for the designer to reduce vulnerability to fire and consequently to reduce fire insurance premiums.

Insurer's technical guidelines

The whole matter of fire and building design is complex. From inception, or acceptance of the brief from the client, the designer of the building will be greatly influenced by the planning laws, Building Regulations and local authority acts and bylaws. The client's attitude to the loss and disruption that result from the effect of a fire in the premises must be considered. This consideration will be influenced by the insurers' attitude to the costs related to damage of the building, its contents or the interruption of the business. The initial design of the building, its fire resistance and the size of its compartments will have a long-term effect on the levels of insurance premium. The presence or absence of a sprinkler system or other active fire protection system will also have a considerable effect on potential losses, and thus on premiums.

Prior to 1986 the tariff insurance companies had founded and subsequently supported the Fire Offices' Committee (FOC). The FOC had rules that primarily affected the fire rating of a building. These rules set down the limits and requirements of each structural element in a building to enable it to conform to a particular standard. The standard was then a guide to the premium rating for fire insurance of the building. A designer, by making a small change in construction, could change a building from a lower to a higher standard, thereby reducing the risk of damage from fire and consequentially lowering the insurance premium rate.

These rules had always been very well respected by almost the entire fire insurance market. They were used as a fundamental fire insurance basis for the whole underwriting world including many non-tariff companies both in the UK and from abroad. These rules were not part of the Building Regulations or Acts or any other statutory requirements. They formed one of the basic considerations in determining the premium rate to be charged on a building for fire insurance.

Since the FOC ceased to be effective in 1986, each insurer is free to set its own premium rates in relation to each risk. Current insurance market conditions result in highly competitive premiums being set, but generally long-term insurance costs for the risk will be lower for more effective building design and fire protection measures.

The FOC technical activities, such as the production of the grades or standards of construction and the rules for automatic sprinkler installations, became one of responsibilities of the Loss Prevention Council (LPC), which was established in 1986 to take over not only these responsibilities but also the test laboratories at Borehamwood, run by the Fire Insurers' Research and Testing Organisation (FIRTO), the Fire Protection Association (FPA) and the Insurers'

Technical Bureau (ITB). The LPC thus became the technical advisor to the insurers, not only those of the old tariff companies but all those within the Association of British Insurers (ABI), together with the Lloyd's Syndicates as well. As a whole, insurers now look to the LPC for technical guidance and refer to LPC publications with regard to the reduction of fire loss in buildings. Designers should be fully aware of the LPC recommended documents at the initial design stage, as the fire hazard of the building will be rated by the underwriters according to important factors such as type of construction, nature of occupancy and management. Premium savings are often lost on the drawing board because the LPC guidance is not considered in the design stage, and by the time the insurers review the fire risk on the building, necessary alterations prove impossible or very expensive to implement.

The LPC represents insurers on national and international standards committees and thus seeks to influence in the first case the requirements of these standards. It also strongly represents the insurers' views to the Department of the Environment, which is responsible for the production of the Building Regulations and to a large extent has seen insurers' requirements written into the base documents, so that there is little contention between insurers' requirements and legislative requirements with regard to fire. With regard to insurers' requirements, five main areas should be considered:

- building construction standards
- rules for automatic sprinkler installations
- rules for automatic fire alarm installations
- fire doors
- code of practice for buildings under construction.

The current guide with regard to construction of buildings for legislative purposes is Approved Document B of the Building Regulations 1991. Despite this including a statement that 'insurers will require their own higher standards with regard to property loss', Approved Document B is generally acceptable to insurers. It is important to recognise that Approved Document B provides a means, but not the only means, of meeting the requirements of the Building Regulations. It is divided into five sections, covering:

- means of escape
- internal fire spread (lining)
- internal fire spread (structure)
- external fire spread
- access and facilities for the fire service.

It is largely a prescriptive document but it makes room for alternative approaches under the general heading of fire safety engineering. As such it makes room for a new direction in building design, where the architect becomes responsible for demonstrating that the building meets the requirements of the Building Regulations based on justification of design that meets the objectives of the Building Regulations in ways other than those normally prescribed. The clause says 'a fire safety engineering approach that takes into account the total fire safety package can provide an alternative approach to fire safety'. It may be the only viable way to achieve a satisfactory standard of fire safety in some large and complex buildings. A British Standard now exists for development on fire safety engineering detailing means by which fire safety objectives can be met.

The LPC publication, *Code of Practice for the Construction of Buildings*, sets higher standards for a number of building features that go beyond the requirements of Approved Document B. Examples would be:

- linings
- the use of thermoplastic materials in windows and roof lights
- the construction of fire break walls
- guidance on protection of structural steel and cast iron beams to achieve appropriate fire resistance requirements.

This document should be taken into account by designers at the early stages of design and the extra provisions in terms of fire resistance required by this document should be weighed against cost and other factors.

Requirements for the design of automatic sprinkler installation

In some buildings there is a statutory requirement to install sprinklers either under the Building Regulations, or through local authority

Acts. The types of buildings where this is applied are offices, shops, assembly and recreation and industrial buildings, as well as storage areas more than 30 metres in height. The Regulations also allow a trade-off in terms of fire resistance with sprinklers installed in buildings of lower height. Occupants of other buildings in which sprinklers are not required by law and are not provided may find it expensive in arranging fire insurance cover. In such cases, the capital cost of installing sprinklers may well be offset by discounts on the fire insurance premiums.

An automatic sprinkler installation provides:

- detection
- warning or alarm
- fire containment
- extinction in most cases.

The FOC rules, which largely controlled sprinkler installations in the UK, have been superseded by the British Standard and LPC Rules for Automatic Sprinkler Installations, which were published in 1990 and have subsequently been added to by technical bulletins produced by the Loss Prevention Council. These all specify the minimum standards required in the installation of sprinklers in buildings under the following general classifications:

- *Light hazard*. This usually covers non-industrial occupancies such as houses, flats, hotels, hospitals, museums and schools.
- *Ordinary hazard*. This covers industrial and commercial occupancies involving the use of materials unlikely to develop instantly burning fires in the initial stages. This classification is divided into four groups, according to the difficulty of achieving fire control with sprinklers. Each group has the same requirement for water density, but as the fire risk increases so the water provision requirement increases to provide for the operation of more sprinklers.
- *High hazard*. these are industrial occupancies with exceptionally high fire loads, either by virtue of special piled storage or because of the presence of materials with potentially rapid burning rates.

A new innovation in the BS 5306/LPC Rules is the provision of specific requirements for life safety. While it is true that sprinkler protection has always provided a high level of fire safety for the occupants, additional requirements for life safety ensure that a higher degree of reliability is achieved. This is done by a requirement for the modification of alarm valves so that they cannot be shut without an alarm being given, and also the provision of by-pass valves so that the system alarm valves can be maintained without the loss of protection. Another feature which has become important is the use of fast response sprinkler heads, which react quicker at the onset of fire; thus the amount of damage and smoke produced from the fire will be less.

Designers should study the detailed requirements of the LPC recommendations at an early stage and take advice from insurers with regard to the appropriate classification for protection. Other areas for consideration should be:

- legal requirements;
- insurers' requirements;
- availability of suitable water supplies to give correct pressure and flow rate;
- effect of the saving on the fire insurance premium related to:
  - building structure
  - plant
  - raw material stock
  - manufactured stock
  - consequential loss policy;
- possibility of writing off the capital cost of sprinkler installation as plant against profits in the first year of installation;
- effect of an extension or alteration to an existing building (unless an extension can be completely isolated from existing parts by 'fire break' floors and walls, in which case the new and old will be rated as one risk based on the worst condition that prevails).

Rules for automatic fire alarm installations

The insurers' requirements for automatic fire alarm installations are included in the British Standard, BS 5839, which provides for two levels of protection, namely:

- life safety protection
- property protection.

Usually insurers' discounts are only given for fire alarm systems directly connected to the fire brigade, thus ensuring reliable response in the event of a fire. Much fire safety legislation under the Fire Precautions Act requires the use of automatic fire alarm systems and these have become very widely used and generally are of high reliability.

The British Standard applies to heat and smoke detectors together with beam and infra-red detectors. It covers such matters as definition, classification of installations, areas to be protected, siting and spacing, and apparatus generally, including location, power supplies, wiring and connection, testing, maintenance and alarms.

The advantages for automatic fire alarm installations will be obvious to all designers in their early consideration of this matter and will be in line with recommendations for sprinklers. There is no doubt that the best warning of fire is the early warning. Efficiency of the fire brigade is not judged by the gallons of water it can pump in a minute but by the minutes it takes for the fire fighting appliance to reach the scene of the fire. Notwithstanding this, the fire alarm system should ensure ease of evacuation of the occupants from the building prior to the arrival of the fire brigade.

The designer has available a variety of warning systems that can be installed in either new or existing buildings in order to detect a fire. Detailed examination of these installations has been made in Chapter 4. The warning systems can alert the occupants of the building and also the fire brigade. Fire detectors are designed to take one or more of three characteristics of a fire, namely smoke, rise of temperature or flame. No single type of detector is the most suitable for all applications and the final choice does depend on individual circumstances. In some premises it may be useful to combine different types of detectors or to install both a detection system and an extinguishing system such as sprinklers. In other circumstances, statutory obligations may require the designer to install alarm or warning installations before a Fire Certificate can be issued.

Warning can be given either manually (in a small building) or automatically. The first reaction to fire is for somebody to shout 'fire'. In many smaller types of buildings this is acceptable as the best method of warning. Fire alarms that can be manually activated by the occupants range from the hand-bell to the push button types that sound alarms throughout the building and raise direct assistance from the fire brigade (manual call points). Flashing lights which are also manually activated are more appropriate warnings for deaf persons; where the designer is concerned with disabled people he must give very special attention to the provision of special visual or audible warnings. The LPC provides a list of approved automatic sprinkler equipment and fire alarm systems, together with the names and addresses of suppliers and approved installers.

Fire break doors, lobbies and shutters

The LPC requirements for construction and installation of fire-break doors, lobbies and shutters go beyond the straight testing to BS 476 to ensure that the component is effective in its installed position. Therefore the LPC requirements not only relate to the testing of the door but also to the means and quality of the installation. In addition to the fire testing of the doors, these requirements include an examination of the durability of the door hardware and the resistance of the door itself to impact.

Construction sites

Significant fire losses occur on construction sites and in buildings undergoing renovation. Architects and designers should be well aware that while buildings are under construction or being renovated potential of fire loss is often greater than when the building is in use. Accordingly the LPC in conjunction with others has produced a joint Code of Practice on the subject. The code specifies attention to fire safety that needs to be given during the design phase and actions to be taken during the construction phase. It details emergency procedures and fire protection measures, and deals with such matters as temporary covering materials, fire extinguishers and security against arson. It also covers the use of temporary buildings on site and the site storage of flammable liquid. It is generally a requirement of the insurers that the Code of Practice is strictly complied with during the construction period of a building if insurance cover is to be given.

Architects beware!

It is not the intention of the author to detail the basis on how a building is rated for fire insurance purposes, but to illustrate that there is a very definite and exacting procedure that is followed by the insurer for this purpose. Architects should make themselves aware of this procedure by obtaining copies of the proper documents and by consulting insurers, so that consideration is given to this matter at early planning and design stage.

The designer must carefully weigh up the advantages and costs of escape against security; compartmentation against open plan; ventilation ducting and suspended false ceilings against providing a vehicle for spread of fire and smoke. He must think 'FIRE' at design stage and ensure that the client gets not only the best return on his capital but also a building with a low fire risk and low fire insurance premium.

'. . . maintenance of escape routes. . .'

Chapter 6

Occupancy and the Fire Certificate

Occupancy of a building

It is unlawful to occupy a new building unless it complies with the requirements of the Building Regulations 1991 and the Fire Precautions Act 1971. These requirements may also apply to old buildings and, in particular, old buildings which have been altered and refurbished. The requirements of the Building Regulations are under the authority of the local building control and the requirements of the Fire Precautions Act are under the authority of the local fire brigade. Where appropriate to meet these requirements, the fire officer will issue a fire certificate and in these circumstances the building cannot be occupied until a fire certificate has been issued or at least an application for a certificate has been refused.

The Fire Precautions Act 1971 is an enabling act empowering the Secretary of State for the Environment to issue Orders detailing the precise requirements for the designated use of a building. This can be a continuing empowerment which could eventually cover the designated use of all buildings. The fire certificate is the weapon of enforcement of the Act.

Whereas buildings with a designated use as hotels, boarding houses, factories, offices, shops and railway premises are clearly defined in an Order, there are others which need special consideration. Premises that are not necessarily permanent buildings, such as moored and beached craft, circus big tops and similar moveable structures, will need a fire certificate before occupation if used for a purpose where the public may be at risk.

Fire safety engineering

The design of each new building should consider the risk from fire by applying fire safety engineering to take into account the following factors:

- the anticipated risk of fire occurring;
- the anticipated fire severity;
- the ability of a structure to resist the spread of fire and smoke;
- the danger to persons in and around the building at risk.

Current Building Regulations require detailed consideration in this connection on the following measures to be incorporated where appropriate:

- adequacy of means to prevent fire;
- early fire warning by an automatic detection and warning system;
- adequacy of means of escape;
- provision of smoke control;
- control of the rate of growth of a fire;
- adequacy of the structure to resist the effects of fire;
- fire containment;
- fire separation between buildings or parts of buildings;
- active measures for fire extinguishment or control;
- facilities to assist the fire service;
- staff training in fire safety and fire evacuation routines;
- maintenance of installed detection, warning and containment systems.

All these factors should be related in design and construction to achieve a very quick exit time of all personnel from a building within three minutes and ensure that smoke is prevented from reaching exit points in not less than five minutes from the commencement of a fire.

There is a cost-effective solution to implementing the requirements of the Building Regulations and the Fire Precautions Act. These solutions should be the basis of very early design consideration and in large complex projects, consultants who specialise in fire safety engineering may be required to produce a total costed fire safety package.

The shared risk

The parties that share the risk leading to the occupation of a building and being in possession of a fire certificate are:

- *Owner; occupier; insured; designer.* They are required to implement the requirements of the Building Regulations and the Fire Precautions Act at an affordable cost; to ensure a continuing maintenance of all detection, warning and containment systems and evacuation routines; and to obtain and display a valid fire certificate and to place adequate fire risk with insurers.
- *Building control.* They must ensure that the requirements of the Building Regulations are properly reflected in the design of the premises and in the construction of the building.
- *Fire officer.* He/she is to ensure that the requirements of the Building Regulations have been approved by building control and that the measures laid down by the Fire Precautions Act have been properly executed in such a manner to enable a fire certificate to be issued. This Act may impose a statutory duty on occupiers to provide additional measures to the Regulations, particularly in respect of means of escape in case of fire.
- *Insurers.* They are expected to satisfy themselves that the building can be properly classified as an insurable risk at a premium rate comparable to its use and construction, with an adequate sum to indemnify material damage arising from the perils insured. Insurers will seek their own standards before accepting the insurance risk and these standards may be higher than those required under Regulations and Acts.

The parties to these risks, liabilities, obligations and benefits should meet at an early stage of design conception and resolve all matters, which will lead to the issue of a fire certificate and occupation of the building. Where appropriate, the meeting should include a consultant in fire safety engineering and a representative of the LPC. Apart from deciding on the measures to ensure the issue of the fire certificate, the meeting should also agree a well-defined programme. Practical Completion Certificate issued by the designer may not in itself allow occupation of the building. The fire officer may be content to indicate that a fire certificate will not be refused and will be issued within a set period, hence allowing occupation. However, in complex situations, such as shopping malls, the fire officer will no doubt state that no occupation can take place without the certificate.

Building Regulations

The basic principles to achieve the issue of the fire certificate are contained within the Building Regulations. There are different obligations relating to fire safety for each type of building designated, but the main headings are as follows:

(1) That there is a satisfactory standard of means of escape for persons in the event of fire in a building. The requirement demands that the building shall be designed and constructed so that there are means of escape in case of fire from the building to a place of safety outside the building, which is capable of being safely and effectively used at all material times. This requirement would not apply to prisons.

(2) That fire spread over the internal linings of buildings is inhibited. The requirement applies to all materials lining any partition, wall, ceiling or other internal structure, and demands that linings shall:
 (a) resist the spread of flames over their surfaces;
 (b) have, if ignited, a rate of heat release which is reasonable in the circumstances.

(3) To ensure the stability of buildings in the event of fire; to ensure that there is a sufficient degree of fire separation within buildings and between adjoining buildings; and to inhibit the unseen spread of fire and smoke in concealed spaces in buildings. The requirement demands that:
 (a) the building shall be designed and constructed so that, in the event of fire, its stability will be maintained for a reasonable period;
 (b) a wall common to two or more buildings shall be designed and constructed so that it resists the spread of fire between those buildings (for the purpose of this subparagraph, a house in a terrace and a semidetached house are each to be treated as a separate building);

(c) to inhibit the spread of fire within the building, it shall be subdivided with fire-resisting construction to an extent appropriate to the size and intended use of the building – this requirement would not apply to prisons;

(d) the building shall be designed and constructed so that the unseen spread of fire and smoke within concealed spaces in its structure and fabric is inhibited.

(4) That external walls and roofs have adequate resistance to the spread of fire in the external envelope, and that spread of fire from one building to another is restricted. The requirement demands that:

(a) the external walls of the building shall resist the spread of fire over the walls and from one building to another, having regard to the height, use and position of the building;

(b) the roof the building shall resist the spread of fire over the roof and from one building to another, having regard to the use and position of the building.

(5) To ensure satisfactory access for fire appliances to buildings and the provision of facilities to buildings to assist fire fighters in the saving of life of people in and around buildings. The requirement demands that:

(a) the building shall be designed and constructed so as to provide facilities to assist fire fighters in the protection of life;

(b) provision shall be made within the site of the building to enable fire appliances to gain access to the building.

Issue of a fire certificate

On receipt of the application, the fire authority may request further information. In particular, they may ask for simple outline plans showing the essential features of the building. The fire authority will then inspect the premises and, if satisfied, must issue a fire certificate. If the authority is not satisfied, a notice will be served on the applicant informing what needs to be done before a certificate can be issued and will specify a time limit for the works to be carried out. If the work is not done in that time, subject to further time allowances or appeals, the application will be deemed to have been refused and it will be unlawful to use the premises as specified.

An example of a fire certificate is included in Appendix 3. In principle it will specify:

- the particular use or uses of the premises that it covers;
- the means of escape in case of fire (a plan will often be used for this purpose);
- the means for ensuring that escape routes can be safely and effectively used at all material times (this would cover such matters, where appropriate, as emergency lighting, direction signs and smoke stopping);
- the means for fighting fire for use by persons in the building;
- the means for giving warning in the event of fire.

The fire certificate may also be used to impose requirements related to:

- the maintenance of escape routes and keeping them free from obstruction;
- the maintenance of other fire precautions specified in the certificate;
- the training of members of staff and the keeping of appropriate records;
- the limitation of the number of persons who may be in the premises at any one time;
- any other relevant fire precautions.

Phased completion contracts

Assuming that a building contract involving phased completion also involves phased occupation, it follows that a fire certificate will have to be applied for and issued for each completed phase before it can be put to use. Thus when planning the building it would be sensible to ensure that each phase complies with the fire authority's requirements and that a certificate is likely to be granted for that part of the building.

This situation would be a very important consideration at the early meeting of the parties concerned and may have an effect on the design, the construction and determining the phases to be adopted.

Alterations and extensions to buildings

Where alterations or extensions are made to existing buildings, the occupier will need to advise the fire officer accordingly. To comply with the provisions of the Fire Precautions Act 1971, the application to the fire officer for an amended certificate should be made before building work starts. There is a close liaison between building control officers and fire officers and quite usually the submission of proposals to alter or extend will alert all parties. However, this is not always foolproof. A simple change of use of the building or a small alteration to a fire or exit door may have a profound effect of the danger from fire. Where alterations or extensions occur, a revised fire certificate is necessary and the onus of procuring that certificate lies with the occupier.

Although the professional consultants may assist the occupier in obtaining or amending a fire certificate, legislation places the responsibility for holding and maintaining a valid certificate with the occupier. The certificate not only covers the compliance with building regulations, but also the use and management of the property. For these reasons, professional consultants should assist but not take the responsibility for applying for, obtaining and maintaining the requirements of fire certificates. Consultants should, however, advise client occupiers to apply for and obtain a fire certificate before alteration work commences and before the property is used.

The lack of a valid fire certificate may provoke insurers to query or reject an insurance claim arising from fire damage.

Validity period

A fire certificate remains valid indefinitely, so long as there are no changes in the circumstances. The onus to notify changes in personnel employed or re-arrangement or alterations to the premises, including any change of use, remains with the occupier (or owner when in multiple occupation). The fire authority has the power to inspect under the Act but there is no provision for regular inspections to take place. The frequency of inspections is determined by the fire authority and varies throughout the country.

Maintenance of means of escape

Continuing maintenance of the means of escape is the responsibility of the building owner and/or occupier and penalties can be imposed by the courts in cases of default. Maintenance requirements are usually set out in the certificate. A typical requirement in this respect would be:

> 'The lives of persons can be seriously prejudiced if, in fire conditions, for instance, the protection afforded to escape routes is damaged, or fire resisting self-closing doors are fastened back in the open position or are wedged open and do not close tightly against their frames.
>
> Applicants acting on behalf of a client should ensure that:
>
> - the owner and/or occupier is made fully aware of his responsibility in this respect;
> - all occupants of the building are fully acquainted with the routes of escape and final exits;
> - the owner and/or occupier establishes trained key personnel on each storey of the building with a view to the speedy marshalling of the occupants and the orderly evacuation of the various storeys should the need arise.'

Fire risk assessment

Legislation can be changed at any time and this particularly applies to fire-related matters. Whatever the outcome of legislation, it is the fire risk assessment that needs to be monitored based upon sound principles and implemented to all new and existing buildings regardless of size, use and location. Remember to notify the fire officer of your detailed intentions before work to new or existing buildings is commenced.

'a dry riser'

Chapter 7
Management

Management disasters

There have been a number of major disasters that have drawn attention, not only to the need for buildings to be adequately designed and protected against fire, but also to the responsibilities of the designer and the manager of the building in relation to fire safety. In particular they have focused on the role of the building manager in maintaining a safe building.

In the twentieth century there have been many major disasters, but three occurred over a relatively short period:

- The Summerland Leisure Complex, Isle of Man, 1973
- The Fairfield Home, Nottinghamshire, 1974
- Woolworth store, Manchester, 1979

All were the subject of enquiries – the reports of the subsequent findings should be compulsory reading for all designers and building managers. What all three fires appear to have in common is that in each case the tragedy arose from the coincidence of a number of factors rather than a single cause; from an accumulation of individual errors and failings rather than from the shortcomings of any one person.

Lessons from disasters

The major lessons to be learned from these fire disasters are:

- the speed at which fires can develop: following the Woolworth fire, tests carried out at Cardington on a stack of packing materials, soft furnishings and bedding showed that the fire would have reached maximum severity two minutes after ignition;
- the rapidity at which smoke can spread, making evacuation difficult and hazardous;
- the fact that the majority of deaths in fires are caused by smoke: of the 18 deaths at the Fairfield Home 17 were caused by carbon monoxide poisoning from smoke;
- the need to maintain the integrity of escape routes: the doors at the foot of the north-east service staircase at Summerland were fitted with panic bolts, but were made ineffective by being padlocked and by there being a car parked immediately outside.

Thus the priorities for management should be:

- elimination of portable heating appliances;
- tidiness in the use of the building and external areas;
- maintenance of the building with particular attention to service installations;
- regular maintenance of the warning system;
- regular maintenance of the direction and escape signs and lighting thereto;
- regular maintenance of fire appliances;
- maintenance of unimpeded escape routes;
- training of staff in evacuation procedures;
- training of staff in first aid fire fighting.

Risk management

Risk management has assumed a high profile. Statistics suggest that 70% of fires relate to poor management and maintenance rather than the nature of the occupying trade or use. Insurers may require warranties to be undertaken to enforce and maintain standards that, if not enforced, render the policy at risk.

Past claims experience will be a good indication of management and their attitude. Insurers will set premium penalties where several claims are made even though each may be relatively small. A greater awareness of risk management will enable the insured to obtain lower insurance premiums by adopting a proactive attitude to identifying and reducing loss potential.

The role of management

The designer must have an understanding of the needs and problems of management. The manager in turn must have a detailed knowledge and understanding of the building, its services and its safeguards, to assist not only in the day-to-day running of the building for its primary use but also to help in any emergency that may arise.

The broad responsibilities of management are:

- the day-to-day operation and maintenance of the building;
- the security of the building;
- the maintenance of adequate safeguards in the event of an emergency.

Present day buildings are complex. The variety of building types with which the designer and the manager are likely to be involved is greater than ever before. Both should be fully aware of the causes and behaviour of fire and of how people are likely to behave in a fire.

At the early planning stage the designer should be able to draw on his knowledge and expertise to try to visualise how fire is likely to affect the building and its occupants, plan to confine the fire within containable areas and provide for easily identifiable routes of escape.

In many cases the client may well be the manager of the building or the manager may be involved from the outset. However, it is quite likely that the manager will be appointed after the design process is complete and will not always be aware of the reasons for the various design decisions.

On completion, the building will comply with the current regulations and the requirements of the fire authority. In many instances a fire certificate will have been obtained. On occupation, management will become responsible for the day-to-day maintenance and assume responsibility for complying with the requirements of the fire certificate.

Probably the most important management task will be to maintain the integrity of escape routes and to ensure adequate smoke-free means of escape from all parts of the building direct to the open air. In a properly designed and maintained building, occupants, including any who are disabled, should have no trouble in leaving quickly and safely. If not properly maintained, these 'safe' routes may themselves become death traps.

The main hazards in escape routes are normally the doors, as these are often misused. The commonest door is the self-closing type and, used as provided, no problems should arise. But so often doors of this type can be found propped open to allow free access during the normal use of the building. Temporary wedges can often become permanent and in some buildings it is not unusual to find such doors fitted with cabin hooks to keep them open.

Corridors leading to protected staircases are often used for temporary storage purposes; usually furniture but sometimes files and other bulky items. The result is at best to reduce the width of the corridor and at worst to block the means of escape altogether.

Potential storage areas may also be found at the foot of the escape stairs; not only can carelessly stacked items block the exit doors but they may constitute a high fire risk, turning a clear route to the outside into a trap.

Exit doors fitted with simple panic bolts or latches are often secured with chains and padlocks to prevent illegal entry. In such circumstances they will also prevent egress in an emergency.

Outside the building, access routes for fire appliances must be maintained unobstructed and not allowed to become car parking areas or stacking places for surplus materials or plant.

The tragedy is that the effect of such misuse is rarely obvious until an emergency arises.

Security

One of the biggest problems facing the manager is that of maintaining adequate security. On the one hand is the need to ensure that the building can be evacuated safely; on the other hand is the need to prevent entry to the building by unauthorised persons. Conversely, special provisions must be made in buildings such as prisons and mental institutions, where containment of inmates remains a factor even in the event of fire.

Crime is a particular problem. Obvious high-risk buildings, such as banks and similar institutions, are now so well protected that robbery has turned its attention to more vulnerable areas, in particular the picking-up points for cash at supermarkets, dairies and shopping precincts. The positioning of secure rooms for the purpose of cashing up, making up wages and the storage of high-value goods needs to be carefully considered with both robbery and means of escape for the staff in mind. In purpose-designed buildings this should be relatively simple. Where problems are bound to arise is in the fitting out of building shells or in subsequent alterations.

Ram raiding, using heavy vehicles to knock holes in external walls or cladding in order to gain access to valuable products, is a risk that has to be countered at the design stage of the building. Tobacco, alcohol, drugs and similar products, stored on or near to the external wall of the building, give the raiders a clear advantage in tems of time needed to get in and out before alarms bring the police or security personnel to the site. The valuable products need to be contained in steel container-type protection situated the maximum distance available from external entry and exit to the building.

Vandalism has to be considered, especially arson. This is difficult to prevent and it is significant that nearly one-third of all fires result from arson. The fire at Summerland, with the loss of 50 lives, was started by boys setting fire to part of the disused fibreglass kiosk that had been left lying on its side next to one of the external walls. Proper maintenance would have eliminated this opportunity and the disaster might never have occurred.

The resolution of the seemingly incompatible problems of emergency escape and security will best be achieved if entrances, exits and security areas are considered at the earliest stage. Entrances should be kept to the minimum necessary for the effective use of the building and emergency exits should be planned to be used solely for that purpose, with suitable quick release security locks. Dual use entrances/exits should be kept to a minimum, in which case the problem of achieving control of the entrances in use and the security of the building when closed will be much reduced.

As well as understanding the building and making sure that staff are trained in fire routines and evacuation procedures, management must also ensure that regular checks and maintenance are carried out on mechanical aids and fire fighting appliances, whether fixed or portable, and the results entered in a manual as a continuous record.

The building manual

The task of management will be greatly helped if all the information relating to the building is available for reference. The initial responsibility for preparing this information ought to rest with the designer. Probably the best way to bring together the diverse information necessary for the effective management of a building is to prepare a building manual or log, which will eventually chart the history of the building from its initial design to its ultimate demolition. This manual, as issued on completion of the building, should contain the client's brief and full information relating to the building and eventually a complete history of all adaptations.

An important part of the manual, usually required by the terms of a fire certificate, is the part covering fire safety and means of escape. These aspects can probably best be illustrated graphically using plans of the building.

Site/location plans should show:

- vehicle access for cars, lorries, service and emergency vehicles;
- emergency escape routes;
- hard standings for emergency vehicles;
- high risk stores for petrol and other flammable liquids, gas and liquid gas, toxic materials, etc.

Plans for each floor should show:

- services, identifying positions and runs of water, gas, electricity and drainage, with all controls clearly indicated;
- high fire risk areas, such as boiler rooms, kitchens, laboratories and storage areas, identifying flammable and toxic materials and switch rooms;
- emergency generators;

- emergency secondary lighting, identifying power source and type;
- escape routes, identifying all fire, smoke and escape doors with direction of opening and types of ironmongery, all positions and wording of signs;
- lifts, identifying those for fire brigade use;
- fire alarm systems, whether manual, automatic, local or linked to the public emergency services;
- fixed fire fighting installations, including smoke/heat detectors, dry and/or wet risers, sprinklers, hose reels;
- portable fire fighting appliances, indicating: positions and types, tools and extinguishers.

From hand-over, the responsibility for the upkeep of the manual will rest with the manager, who should ensure regular monitoring and updating of:

- plans, ensuring that they incorporate any changes in structure, layout, usage, fixed and portable equipment and services;
- fire drills, noting frequency with any observations;
- equipment tests, ensuring that all fixed and portable equipment and installations are serviceable;
- training, ensuring that staff are fully conversant with fire fighting and evacuation procedures;
- all necessary licences and certificates.

The building manual is an important document and, together with the fire certificate, should be kept secure but accessible. The designer should pay particular attention to the preparation of the manual and should impress on the manager its value as a management tool.

The building manual should be regarded as an independent but supportive document to the requirements of the current CDM Regulations.

Alterations

Buildings change hands from time to time and very often their use may change. Change usually brings about alterations, from the *ad hoc* alterations made by the works manager to cater for a local change in use, to total refurbishment. If any alterations are attempted without a knowledge of the original brief and design philosophy, potential hazards may be brought into the building that may not become apparent until disaster strikes. Indiscriminate alterations without a full understanding of the possible implications of such modifications can produce potentially dangerous situations, particularly if these are attempted without the assistance of qualified professionals.

No alterations, however innocuous they may seem, should be made without first advising the fire officer, and consulting an experienced fire prevention or safety officer and, equally importantly, the insurance company. Alterations to the building may involve the shutting down of a sprinkler system, which will require a dispensation to be obtained from the insurers for the period involved. In any case the fire certificate must be amended or a new certificate issued if required.

Maintenance

Repairs and replacement of materials, particularly where they affect means of escape, must be of the same grade. Fire-resisting glazing must be made good to the original standard, intumescent ventilating grilles and smoke seals properly replaced and flame retardent linings and curtains maintained.

Fire fighting

Designers should be aware that, while staff will not be expected to cope with fires beyond the early stages, they will be responsible for the safe evacuation of the building. Therefore the correct positioning of static hose reels and portable appliances, together with easily identifiable escape routes, will be vital. The regular fire brigades must not be relied on to assist in the evacuation – they may be involved with another fire or disaster, or their access may be impeded elsewhere.

Fire assembly points

Fire assembly points should always be defined in an open external area well away from each building, with clear sign writing in large letters. Procedures should be in place to know at all times the number of personnel inside buildings to be accounted for in an emergency evacuation. Appointed fire control stewards within the staff should have distinctive overclothing, should position themselves initially at fire exits and should have immediate access to the records to ensure that all personnel are assembled at the external fire points within a prescribed time from the sounding of the alarm.

Locating the fire

When the fire brigade arrives the source of the fire can usually be readily identified in the case of a small or simple building. In a large building, or in a building complex, it will be helpful to have a zone locator, together with a set of key floor plans at every entrance, in order that the fire chief can quickly identify the seat of the fire. Speed is the essence of fire fighting and the ability to be able to pinpoint the source of the fire and deploy resources quickly and effectively may make all the difference between local damage and total loss. It will be helpful to deposit a set of plans with the local fire brigade. These should show a key identifying the various areas. The plans must be kept up-to-date or their value will be lost.

This is particularly important in the massive complexes that are increasingly common today:

- where access may be complicated by separation of vehicular and pedestrian traffic;
- where large volumes are involved;
- where large numbers of people are working or at recreation;
- where the use to which space is put is frequently changed.

The fire brigade will assist in training permanent staff and (of interest to insurers) maintain an actual presence during special occasions.

'.. and whilst we're about it, increase our fire insurance by thirty percent...'
bob

Chapter 8

Sums Insured

Accepting the risk

Once a new building is completed and handed over, or a contract has been signed for the purchase of an existing building, the new owner takes on the responsibility for fire insurance and the type of policy adopted depends on his requirements and attitudes. A landlord or owner-occupier may have some flexibility in choosing the most appropriate form of insurance; a lessee will need to comply with the terms of the lease.

Covering the risk

With rare exceptions, all property insurance is based on the broad principle of indemnity. The insurer underwrites the risk to indemnify the insured up to the value for losses arising from the eventualities set out in the policy. The intention of the policy is to provide indemnification against losses so as to put the insured back into a position comparable to that before the loss occurred.

In what is generally referred to as a fire insurance policy, the prime risk is damage by fire. Fire damage itself can extend beyond simple combustion and covers other heat effects, including melting, distortion, cracking, expansion, contraction, and damage from smoke, water and mechanical reaction. However, the policy may cover a range of specified perils, most often under one contract. This may include damage arising from any or all of the following:

- fire, including earthquake fire, subterranean fire and spontaneous combustion
- smoke
- water to extinguish fire
- lightning
- aircraft
- explosion, other than of boilers or steam pressure plant (insured under separate engineering policy)
- earthquake
- riot, strikes, civil commotion
- malicious damage
- storm or flood
- escape of water from apparatus
- sprinkler leakage
- impact
- subsidence, ground heave, landslip
- theft damage.

Explosion arising from a defective boiler or steam pressure plant is not normally covered by the peril of explosion. It is therefore necessary to complete a separate engineering policy to cover this peril. These policies in addition usually provide for annual inspection of the plant by a qualified engineer.

An alternative to the specified perils policy is the modern commercial all risks policy. It covers all the above perils (subsidence as an option) plus accidental damage. The reader is referred back to Chapter 2 and the law cases of *Austin v. Drewe* (1816) and *Everett v. London Assurance Co.* (1865), and the legal definition of fire. The risk required to be covered by the insured must be clearly defined and properly reflected in the wording of the insurance policy.

Whatever the risks to be covered, one major item to be determined in the policy will be the sum insured. Not only does the sum define the financial limit of the liability of the insurance but it is likely to be used in other negotiations. To ascertain the correct and appropriate sum insured the next major item for consideration is to agree the basis of insurance.

Basis of insurance

Accepting that all insurance provides indemnification for the insured, there is an alternative basis whereby reinstatement clauses can be included in the policy. Consequently, there are alternative methods of assessing the level of compensation for damage suffered, namely:

- indemnity
- reinstatement.

Broadly, the value of a building assessed on an indemnity basis will reflect its age, state of repair and suitability for use, whereas the value assessed on a reinstatement basis will reflect the cost of replacement of that building irrespective of such age, state or suitability. The former will tend to produce a lower figure and the latter a higher one.

Each alternative merits fuller examination and this will follow. First, however, there is one further principle of insurance practice that must be understood: 'average'.

Average

Average is an insurance term and relates the total value of the material asset at risk and the financial loss to the adequacy of the sum insured. Irrespective of the basis of the insurance, the sum should be adequate to meet a total related to the type of cover provided in the policy. Where the sum insured is inadequate, the compensation will be reduced in proportion to the shortfall of the material asset at risk to the sum insured. The formula for compensation will be either:

$$\text{cost of repair} \times \frac{\text{sum insured}}{\text{asset at risk (reinstatement)}}$$

or

$$\text{value of damage} \times \frac{\text{sum insured}}{\text{asset at risk (indemnity)}}$$

An illustration related to reinstatement may make the position clearer by taking the cost of repair of a fire loss at £30 000, the sum insured on a reinstatement basis at £100 000, and the cost of totally reinstating the building at £250 000. The financial risk of the insurers is limited to the sum insured and therefore the compensation would be:

$$£30\,000 \times \frac{100\,000}{250\,000} = £12\,000$$

The £18 000 balance of the repairs would have to be met by the insured.

If the same figures are related to values and indemnity instead of costs and reinstatement, the same situations of average exist. It therefore becomes of great importance to ensure that the sum insured is adequate to meet the total reinstatement cost ('reinstatement') or total value ('indemnity').

Overproviding in the sum insured brings no advantage, except to the insurers by way of increased premium. Under no circumstances can average be operated in a manner to increase the financial compensation. Average is a protection to the insurers to limit their share of the risk by reference to the sum insured.

Indemnity

Indemnity is related to the value of the material asset. In insurance terms it is often assessed by depreciating the current purchase price of labour and materials needed to reconstruct the building by allowing a deduction for wear and tear. Indemnity might therefore be described as the depreciated value of a material asset. If indemnity were applied to a bituminous felt-covered roof, the view could be offered that the roof structure has no depreciation whereas the felt covering should be depreciated in line with its remaining effective life. In a similar way this could apply to decorations, electrical installations, heating systems, floor tiling and other items having a limited life. Brickwork, concrete, steel and timbers should not normally qualify for depreciation unless there is evidence of excessive wear and tear or structural failure.

The application of an indemnity valuation to buildings has to be applied to the whole of the building and not to a part. There is, however, a problem of application that can be simply illustrated. Assume a warehouse has a market value of say £200 000, of which the land is worth £40 000. An indemnity policy to cover the fire risk on a building would therefore set the sum insured at £160 000. If the warehouse were totally destroyed by fire then £160 000 would be the correct sum for payment for indemnity compensation. The land could be sold for £40 000 and, together with the proceeds from the insurance policy, the insured would have a combined total of £200 000. Provided a similar type of warehouse was available and acceptable at that value (£200 000), the owner would then be in a position to carry out the purchase. Hence following a total loss, the indemnity arrangement would have placed the insured in a position similar to that before the loss: a 'like for like' position.

However, it is unusual for damage by fire to result in total destruction. More often a relatively small percentage of the building is damaged and repair has to be undertaken. The indemnity cover now becomes a notional situation as it is impossible to carry out repair with materials and labour that are depreciated by wear and tear. The notional repair costs can only be redressed by a financial contribution, being a proportion of the cost of repair using new materials and labour. Building costs now enter the equation. To rebuild the same warehouse would cost say £300 000 and being insured on an indemnity basis the compensation would be in the ratio of £160 000 to £300 000 of the repair costs. The insured would find that almost half the repair costs were uninsured.

The matter could become very severe if the warehouse were completely destroyed and if, because of trading facilities, it became essential to rebuild on the same site thus providing a new building in lieu of one in a poor condition. Furthermore, the use of the fire-damaged building can also come into the equation. A barracks no longer used for lodging soldiers but currently used for storing surplus farm machinery or goods would have an indemnity, because of its current use, related to the value of a building similar to an open barn.

In *Exchange Theatre Ltd* v. *Iron Trades Mutual Insurance Co. Ltd* (1983), a Victorian hall that was being used by the insured as a bingo hall and club was damaged by fire. The court held that the loss was not the reinstatement cost of the Victorian hall but the value of a modern equivalent building suitable for the use of the insured as a bingo hall.

Betterment

The difference between the indemnity value of the old materials and the cost of new materials is called betterment. Betterment financially quantifies the improvement of new materials used in reinstatement or reconstruction over the depreciated value of the old materials in a building before the fire took place. Provision for the cost of betterment in the case of 'new' for 'old' materials would be made in the sum insured and the cost would be covered in a reinstatement policy.

Betterment can also apply to improvements from redesign or use of more expensive materials. Following damage by fire, the insured may elect to extend the floor area of the building. Alternatively or additionally, the heating may be exchanged for air conditioning or a felt roof covering may be replaced by asphalt. Provision for this type of betterment is neither covered by an indemnity nor a reinstatement policy.

Where a building is covered by indemnity insurance and repairs are undertaken incorporating betterment items, it invariably becomes necessary to estimate the building costs of reinstatement without regard to the betterment. This enables the adjuster to relate the total reinstatement costs over the actual cost of repairs including betterment, as a ratio of the sum insured as indemnity, to give a value for the damage of the betterment items in a similar way to average.

$$\frac{\text{estimated reinstatement costs}}{\text{actual costs including betterment}} \times \text{sum insured (indemnity basis)}$$

Where a building is covered by reinstatement insurance, the betterment is the difference between the actual cost of the item of work and an assessment of the cost of that item if reinstated with areas or materials similar to those existing prior to damage.

'Negative' betterment, by using cheaper materials or smaller areas in reinstatement, is not generally recognised by insurers.

Reinstatement

In determining the sum insured for a reinstatement policy the insured is providing to recover the cost of repair, whether the damage is partial or complete. The advice on this sum should come from a quantity surveyor, as he is clearly the best qualified person to determine likely rebuilding costs. Where the building is new and the quantity surveyor has been concerned with the construction, the details from his final account will provide current and reliable cost information. In cases of existing buildings where no cost information is available, the quantity surveyor will rely on his professional expertise in arriving at an approximate estimate of the rebuilding costs. The reinstatement cost will incorporate many factors, which together will enable the insured to be reimbursed for his total expenditure in repairing or replacing the building after damage by fire. Whatever has to be paid for the repair of the building will be reclaimable from the insurers provided the sum insured gives adequate cover. Unlike the indemnity basis, new materials will replace old without application of betterment.

The first factor for consideration in calculating the reinstatement sum insured is to determine the cost of rebuilding the structure on a prescribed date, with materials and in a style similar to the existing building. The quantity surveyor will use known cost information, building indices, approximate quantities and pricing, floor area or cubic content, estimating techniques related to other known building costs, or any combination of these methods, to arrive at the reinstatement value. The figure at this stage is notional as it will relate to a given date, probably the first day of a new annual insurance contract. The notional cost would assume that the building could be replaced on that day without influence by other factors that, in practice, will always impinge on a notional situation. This notional estimated cost of reinstatement is known as the declared value.

Apart from the declared value of reinstating the building, the valuer must determine whether the sum insured to be stated in the policy automatically provides for such matters as are listed below, or whether it has to be adjusted to take account of them.

- Inflation of building costs during the period of obtaining approvals for planning, building regulations and obtaining competitive tenders.

- Inflation of building costs during the period of site clearance and reinstatement of the building. This would be an average inflation estimate for the rebuilding period.
- Inflation of building costs during the period of the insurance policy, which is usually one year. Where small sums are involved this addition is recommended, but when applied to large sums the inflation element should be provided by an escalation clause in the policy taking into account some recognised cost indices. The RICS House Rebuilding Cost Index is an example. This allows for provision of the full inflation throughout the length of the insurance period by the wording of the policy.
- Costs of demolitions, removal of debris and general site clearance arising from fire damage.
- Costs of protecting, shoring and supporting adjoining property arising from fire damage.
- Cost of temporary works and protection to insured property.
- Costs of complying in the reinstatement with current Building Regulations. Many existing buildings have unprotected steel frames, unlined roofs and uninsulated cavity walls, which would need alternative and more costly treatment in a major repair or reconstruction situation. These costs and those that can be defined relating to planning requirements are usually dealt with in fire insurance reinstatement policies by means of a memorandum or extension to the policy, called the 'local authorities clause'. It is important that these additional costs are included in the sum insured and are covered by a suitable clause in the policy.
- Provision of planning requirements is fundamental to the sum insured. Where special consideration will be required in the reinstatement after material damage by fire, the problem should be discussed and resolved between the insurers and the insured before the appropriate sum insured is determined. This matter is dealt with in detail in Chapter 9.
- Fees payable to building control in respect of submission of plans and inspections.
- Cost of upgrading existing sprinkler installations, water spray and deluge systems. Any reinstatement of these installations and systems will have to comply with current regulations and LPC requirements. Where the existing installation is below the prevailing standards, insurers take the view that the additional cost

of upgrading in these circumstances is not covered by the provision of the local authorities clause in a policy. It therefore becomes necessary to arrange a separate insurance item where this particular risk occurs, or to have a special memorandum written into the policy.

- Costs related to the tendering climate of the construction industry projected through the period of insurance. The sum insured always relates building costs to a future period and to a situation where speed of reinstatement is often vital. Costs established during a period of general recession in the construction industry and where urgency of building does not apply can become misleading in determining the sum to be insured if the market conditions change. The position should be assessed and an allowance made if applicable.
- Costs of professional fees and disbursements in preparing and arranging tender and contract documents for the reinstatement work, and the cost of inspecting the work and preparing the interim certificates and the final account. An enabling contract may be necessary to carry out demolitions, remove debris, shore up adjoining property and make the site protected and safe. Allowance for fees should reflect the likely particularities of the building or site being insured.
- Leasehold premises usually have provision in the lease for insurance of loss of rent arising from fire damage. This additional sum should be shown separately in the report. This is one of the few items of consequential loss that is usually covered in a traditional type of fire policy. It arises because of the wording of the lease – the loss of rent relates directly to the building and not to profit arising from its use.

Value added tax

Value added tax (VAT) in respect of goods, labour and services supplied by the construction industry is currently standard rated except in certain specific circumstances. VAT is part of the Finance Act which is an annual event directly related to and confirming the annual government budget. The law relating to VAT can therefore change at almost any time. The implementation of VAT is vested with HM Customs and Excise.

HM Customs and Excise publish notices from time to time – currently those applicable to the construction industry are detailed in Notice 708. This notice is informative and should be considered in detail by the insured to determine whether or not the standard rate of VAT should be added to a valuation and allowed in the sum insured.

Almost without exception, the sum insured for a building will be in excess of the upper limit of the annual turnover to allow a construction firm to be nonregistered for VAT. The insured must therefore determine with the local Customs and Excise Officer that the insured, due to their own VAT status, can recover the tax that would be charged by the construction industry arising from the future reinstatement of material damage. If not, allowance for VAT should be made at the current rate and included in the sum insured.

The current regulations need careful consideration. There is no Customs and Excise Notice particularly directed to VAT charges arising from fire damage and reinstatement. Notice 708, *Buildings and Construction*, would apply. Unless the building damaged is totally destroyed and complete reconstruction is required, the reinstatement would fall under *Repair and Maintenance*, which in all cases is standard rated.

Whereas certain new buildings, when constructed or totally reconstructed, are zero rated, any partial fire damage reinstatement will be standard rated in all cases. It would be ill advised to consider the sum insured on the basis that future fire damage would result in total reconstruction. Even where the damage represents 90% of the total, the insured may be restricted in choosing total reconstruction by planning requirements, particularly in respect of listed buildings. As long as professional consultants are VAT registered, all professional fees are standard rated whether the building is totally or partially reinstated.

Where the insured is unable by his status to reclaim VAT, then the sum insured should make due allowance for this charge.

'Services below ground'

Chapter 9
Preparing a Valuation Report

Basis of policy

Before a valuation is prepared on a material asset such as a building, it is essential that the basis of the insurance policy is clearly determined and fully understood by the valuer. The provisions regarding rebuilding and indemnity listed in Chapter 8 will clearly be appreciated by a surveyor or architect, but there are insurance arrangements that should also be given full attention in the valuation. By their nature these are insurance matters, but in many instances they affect the detailed work involved in preparing a valuation for fire insurance purposes.

Before preparing a valuation for fire insurance purposes, the best advice is to call for a copy of the proposed policy and read it carefully. Pay particular attention to the basis of the sum insured and the regulations on payment against claims. Always clarify matters with the insurer, confirm any explanations in writing and record them in the body of the report and valuation. The underlying basis of the cover is that the sum insured is notional and related to a certain date (usually the first day of the policy, called *day one*). Where the sum insured is on the basis of day one (*declared value*) it assumes that the building could be repaired, reinstated or reconstructed on day one; also, allowances for inflation during the periods of demolition, site clearance, planning, reconstruction and the actual insurance period are all covered by the wording of the policy and premium payments. Where these allowances are not part of the policy wording, financial allowance will need to be included in the valuation.

Although not necessarily comprehensive, the valuation should cover, and the valuer should determine beyond doubt, the following (where appropriate):

- the correct address of the property;
- a schedule of the buildings and their use;
- the gross internal areas of the property;
- the basis of the sum insured (either indemnity without qualification or indemnity incorporating the reinstatement provision of providing new for old materials);
- the basis of the valuation for reinstatement providing the date from which the base values are derived;
- professional fees and disbursements;
- provision for inflation in building prices;
- VAT;
- provision for loss of rent or cost of alternative accommodation;
- amount of excess, if any;
- items insured under a first loss policy;
- period of insurance;
- scope of reinstatement.

Address

The importance of determining the correct address and extent of the property to be insured and valued is vital. Quite rightly, the valuer is easily led to believe that his clients will know precisely the correct address and extent of their properties. In most cases this is probably true, but surprisingly records often become vague in a large company and mistakes, acquisitions and disposals are not properly rectified or recorded. In addition, some properties have a road number for a ground floor shop and a suffix (a) (b) or (c) for upper floors that may or may not be part of the sum to be insured.

A check should be made that the property being valued is precisely stated on the insurance policy. There can be no golden rule for determining that the address is correctly stated, but extreme caution should be exercised in obtaining information without detailed and proper enquiry.

Schedule of the buildings

Where the valuation is in respect of more than one building on a site with a collective address, each building should be scheduled in the report, providing in each case:

- use or name of the building;
- number of storeys;
- plan reference related to the fire plan prepared by the insurers;
- brief description of construction;
- gross internal floor area storey for storey.

Basis for sum insured

As far as insurance of buildings is concerned, there are two broad bases for the sum insured. One is indemnity and the other is reinstatement. The difference between these two bases is given in Chapter 8. Valuation for indemnity of a property would generally be made by a valuation surveyor reflecting current market price. Valuation for reinstatement relates to construction costs in reinstating the building; this would generally be given by a quantity surveyor.

The valuer should explain the difference to the insured before commencing the valuation, in order to determine that the insurance policy and sum insured meet the requirements of the client.

Basis of valuation for reinstatement

As stated at the start of this chapter, the underlying basis of valuations of buildings for a reinstatement insurance policy will usually be related to a notional level of cost for reconstruction, applied at the inception of the period of insurance. This assessment is termed the 'declared value'.

The declared value should include the cost of the following where appropriate, provided the insured requires the cover in the valuation and that none of the items are insured through other policies.

- all items forming the foundations, structure, finishings, fittings, services and all matters comprising the buildings to be insured;

- external items and main services;
- additional expenditure of reinstatement to comply with public authority requirements;
- demolition, site clearance and removal of debris arising from damage to property;
- shoring to adjoining property arising from damage to the insured property.

The insurance cover for the declared value can be arranged through different types of policies, although the most common in use is called the 'day one' basis. Reinstatement in respect of a day one basis policy means the insured's assessment of the cost of reinstatement of the property including all appropriate items listed above, at the level of costs applying at the inception of the period of insurance. For this purpose reinstatement means:

- the rebuilding or replacement of property lost or destroyed which, provided the liability of the insurer is not increased, may be carried out in any manner suitable to the requirements of the insured on the same site or upon another site;
- the repair or restoration of property damaged, in either case to a condition equivalent to or substantially the same as, but not better or more extensive than, its condition when new.

The valuer should be watchful and take into account in the declared value all matters that will affect the reinstatement cost. Where dealing with older types of building, special consideration should be given to:

- access, which because of subsequent constructions has resulted in new restrictions;
- availability of materials, which may no longer be manufactured;
- planning requirements, which are dealt with in some detail later in this chapter;
- redundant work, which would not be replaced;
- obsolete items, which are not covered by the public authority requirements but would need to be upgraded to meet other requirements, e.g. sprinkler installations.

Where a special condition has a financial effect on the declared value, the valuer should allow for this accordingly. In addition, the report should state the special conditions so that proper disclosure is made to the insurer before the policy is finally drafted.

In preparing the details of a valuation, the valuer should not be misled if tendering is currently very competitive. Reinstatement after material damage often requires urgent action on the site. In many instances there is simply not enough time to have full specifications, drawings and detailed bills of quantities prepared to gain the best financial tender in the market. The basis of the valuation should be related to reasonable building costs, in order to ensure that the sum insured is adequate to meet an 'open book' prime cost situation. More can be read on this point in Chapter 12.

Professional fees

Unless covered by a first loss policy, professional fees and disbursements are part of the declared value and should be included as such. It is better to show this cost separately in the valuation report and relate it to a recommended institution scale. In the event of a claim arising, the insured will need immediate, capable and expert advice. Should the damage and resulting claim be significant, the insured will need to consider:

- the financial claim against the policy, the cost of preparation of which is not reclaimable under the policy;
- the method, speed and quality of the reinstatement.

The experienced professional team should, in most cases, be capable of producing documents in such a manner that the proved reinstatement costs form the basis of the financial claim.

The calculation of the professional fees and disbursements to be included in the declared value should allow adequately for a full service by each of the design team; it becomes an advantage to disclose the extent of this allowance to the loss adjuster by reference to the valuation report. Insurers will attempt to use the indemnity aspect of the policy to impose on the reinstatement the cheapest professional solution; in a claim situation this should be resisted if the scope and quality of the professional services are not to be impaired or restricted.

Inflation

Provision for inflation in building prices must be considered in every sum insured. These provisions cover the increased costs for the following periods:

(1) site clearance, debris removal and any enabling preliminary work;
(2) planning, procurement and site mobilisation;
(3) reconstruction;
(4) the length of the policy.

Should an incident occur on the last effective day of the policy, the total time from the date of inception of the policy to encompass the above periods may be several years. The sum insured must therefore adequately allow for this eventuality by either the wording of the policy or the inclusion of a sufficient additional cost.

Many policies for domestic property have no alternative other than to increase the sum insured to cover inflation in building prices adequately for the periods stated above. There is no declared value in such policies, only the sum insured to cover all costs.

In the case of policies on a day one basis using a declared value, the provision for inflation is covered by the wording of the policy. Each insurer will have their own solution on this matter and will market it to their advantage. The insurance premium will be adjusted to cover this risk. The most common methods of covering the inflation amount are listed below:

- The inclusion of a lump sum in addition to the declared value, where the premium is at a lower figure than that charged on the declared value.
- A percentage stated on the declared value.
- The declared value is linked to a building index allowing the sum insured to adjust accordingly throughout the period of the policy. Great care must be taken to ensure that this type of allowance will continue beyond the last date of the policy, in order to cover the total periods already listed. If this is not so, then a further amount to cover the first three periods listed must be added to the sum insured.

Value added tax

VAT has already been referred to in Chapter 8. The main concern in preparing a valuation for fire insurance purposes is to determine the VAT status of the insured: are they VAT registered or not?

The insured who are VAT registered will have to declare this in their proposal for insurance. The general rule allows any VAT charge by the construction industry to be reclaimed from HM Customs and Excise. This matter, however, is in each case a particular arrangement between the insured and HM Customs and Excise, and each insured should have written confirmation from HM Customs and Excise that VAT can and will be repaid in the circumstances of reinstatement. If these circumstances are clearly defined and confirmed by HM Customs and Excise, then the valuation excludes VAT and the report gives an appropriate statement on this exclusion.

The insured who are not VAT registered will have to declare this in their proposal for insurance. The valuation will include an appropriate sum to cover this charge. Again, the report should give a statement on this inclusion.

Rent

Provision for the loss of rent or the cost of alternative accommodation is another risk to be considered in the valuation report.

Loss of rent is usually related to the financial obligation of a tenant or leaseholder to the landlord. This provision would be in the name of the landlord as the insured and is related to a lump sum representing the current rent receipt for the estimated period of total reinstatement. It would need to be declared as a separate amount in the policy and accordingly shown as a separate sum in the valuation report.

The risk of cost of alternative accommodation can apply to owner-occupiers and tenants. If not allowed in the building policy, it can be provided in the consequential loss policy. The risk in each situation will be determined by the individual circumstances.

Where a lease is in existence both landlord and tenant should be made aware of their obligations and liabilities, and should provide adequate insurance to meet all reasonable eventualities. The lease should be carefully read by the valuer prior to his report. A damaged building may result in the loss of rent to the landlord and alternative accommodation may cost the tenant more than the rent being paid before the damage: two responsibilities; two insured; two insurance covers required.

Excess

Excess is the agreed amount (if any) that the insured will pay for the first part of a loss without being reimbursed by the insurer. The greater the sum of the excess, the greater will be the reduction in the insurance premium.

Where the valuation is for one building, the degree of risk in assessing the amount of excess to be stated in the policy will be with the insured. However, the valuer should be aware of, and may well have recommendations on, excess in respect of a sum insured covering many buildings.

An example would be an industrial complex where there is one sum insured for all the buildings. Perhaps a range of large buildings, each valued in excess of £20 000, and a number of small buildings, each isolated by a considerable distance from each other and from the main range and each valued at less than £20 000. The valuer should consider with the insured the saving in premium on the sum insured where excess is £20 000. It may prove that the saving in premium over three or even two years would be sufficient to cover the risk of loss on any one of the smaller buildings. If the insured decides to accept that risk by use of excess and gives written instructions, then the valuer would exclude the value of those smaller buildings from the valuation report and clearly state the reason.

If the industrial complex is very substantial, such as an oil refinery, the excess could be, say, £100 000 or even £250 000 and the understanding of this arrangement consequently becomes more pertinent to the valuer.

First loss

In simple terms a 'first loss' policy provides for insurance cover for the first part of a possible total loss. For example, a building or buildings may have an estimated total reinstatement cost of £2 000 000 and the insured wishes only to insure on a first loss basis the reinstatement insurance on the first £500 000. Repairs resulting from fire damage up to a total reinstatement cost of £500 000 would be covered by the policy but the remaining £1 500 000 would be uninsured under that policy. Average does not apply to a first loss policy.

There are a number of reasons that the insured might decide on first loss cover only. One reason might be the terms of the lease. Another could be the insurance cover on 100 pig sties dispersed over a large area of land with little risk of all the sties being damaged in one incident.

Where a valuation is required relating to a large risk spread over many properties and relating to a number of different locations, the insurance brokers may advise that a first loss policy should be arranged on certain specific items. Examples of this procedure could apply to sections like demolitions, professional fees and external items such as pavings, boundary walls, landscaping, fencing, gates, floodlighting, sewers and so forth. In these circumstances the valuation should advise on the situation that would provide the largest risk of a specified section and the first loss policy would have a liability related to the value of that particular situation. The largest risk situation for one first loss policy (e.g. demolitions) could be on one site and for another (e.g. fees) could be on a different site. In respect of each specific item or items, the largest risk valuation would become the sum insured for that first loss policy. However, the first loss policy would apply and provide insurance cover on the specific item or items on all the buildings at all the locations.

The valuation report should state which items have been instructed on a first loss principle and then exclude the value of those items from the recommended sum insured for the general reinstatement policy.

The premiums for first loss policies are usually higher than for a general total cover policy.

Period of insurance

Although premiums are usually paid annually, the period of insurance can be for an extended period. Where a long-term agreement is in place this refers to the period of contract between the insured and the insurer. The policy issued is usually for one year, but a discount on the insurance premium is allowed when renewal is guaranteed between the parties for a longer period of time. Long-term agreements can be for any agreed period, but in the interests of both parties the term is usually restricted to three years. Sums insured still need to be reviewed and determined annually.

Scope of reinstatement

A report and valuation on the sum insured should clearly state items and allowances that are included and those that are excluded. This not only defines the scope of the valuation, but is extremely useful for reference and identification in the event of damage by fire. It is essential that plan references should relate to the fire plan prepared by the insurers. Where it is necessary for insurance purposes to know the estimated time required for reinstatement, this should be stated. Although a valuation report for fire insurance purposes may show separate figures for each plan reference, this would be used for rating the insurance premium only. The sum insured should be one figure for the whole location and based on a clear site with reinstatement in one continuous building operation.

Cost of foundations, drainage and services below ground should be included in the calculations, as most serious fires result in the need for their reconstruction. If the foundations are piled, the cost of piling can be excluded from the sum insured as it is unlikely to be affected by fire damage. Very deep foundations may also be excluded, but it is important to specify clearly such exclusions, so that the costs are not used in average calculations and the insured knows that he is totally at risk on those items.

Services need careful consideration in determining whether they are regarded as insurable as part of the building or as plant. Decision of classification in this respect is between the insurer and the insured and often follows the manner in which items are regarded in the firm's accounts. Electrical supplies to machines may be listed as plant or building and likewise similar consideration should be given to lifts, hoists, sprinklers, smoke detectors, boilers, storage tanks, heaters, light fittings and many other service items. The important task is to list all relevant items, then to determine under which heading they are to be insured and to confirm in the report whether the items are included or excluded.

Similarly, a list of particular items should be considered that covers concrete pavings, tarmacadam, factory or estate roads, car and lorry parking, stacking areas, sewers and mains supply under factory roads, fencing, gates, ramps, dock levellers, road lighting, sewage pumps, greenhouses, garden sheds and all other external items. These items may not need insurance cover and it is most important to identify the component items of the premises and to indicate whether they are included or excluded from the sum insured.

In preparing a valuation recommending the sum insured for a building on a reinstatement basis for fire insurance purposes, the valuer must emphasise that the figures have a limited use. They should not be published without written consent in case they are misinterpreted. The report should state that the valuation has no direct relation to market values nor can it be used in respect of current cost accounting for the basis of depreciated replacement value. Situations of fire insurance of large country mansions, whether they are used for their original purpose or in an alternative way, illustrate the need for the surveyor to identify clearly in the report that a valuation for replacement or reinstatement can be an entirely different figure from a valuation for market value.

Planning requirements

Mention has already been made that current planning requirements may influence the surveyor's approach to preparing a valuation for insurance purposes. An existing factory constructed of steel and externally clad with corrugated iron could be permitted to remain in use indefinitely. As soon as it is substantially damaged by fire the planning authority may require an entirely different type of building as replacement. The surveyor should be made aware of the situation, explain the position to the client and provide in the report and valuation for either:

(1) a sum insured on an indemnity basis, related to the market or book value of the factory, which would become reclaimable

from the insurers if the fire damage invoked a notice from the planning authority effectively requiring the damaged building to be totally removed, or

(2) a sum insured on a reinstatement basis, related to the cost of removing the remains of the existing damaged building and replacing it with one acceptable to the planning authority.

The insurance premium would be considerably less in (1) than in (2) for obvious reasons, but the security of the insured would be greatly increased in (2) since finance would be available to rebuild the factory. Additionally, in (1) no proper provision is being made for the situation of a small fire causing damage that would not provoke the planning authority to require total removal. The insured would then have to repair the damage and the indemnity insurance would prove insufficient due to average, for the reasons described in Chapter 8.

Where a compulsory purchase order has been served in respect of a building and it has been agreed by the authority that demolition will take place on the expiry of the order, the insurance provision for the unexpired period is best arranged on indemnity. In the event of fire, the insured will have to admit that compensation in accordance with the order will be received and that reinstatement is out of the question. In this situation the sum insured for the indemnity insurance should be related to the rental value of the building for the unexpired period of the order.

A further consideration is the situation of a building being used in a different way to the general zoning laid down by the planning authority. A factory in an area planned for residential use provides an example. So long as the factory is not served with a compulsory purchase order, the building would continue in use with existing user permission. In the event of serious fire damage, permission to rebuild may be refused. In these circumstances the insurance should be arranged for reinstatement on an alternative site and the sum insured related accordingly to the anticipated costs. In such a case several items that may be excluded would now form part of the sum insured, including piling, concrete access roads, hardcore stacking areas, main services, sewers, general site layout items and even consideration for differing land values. A policy to cover this type of risk would need very detailed negotiation with insurers.

Listed buildings should have special consideration in determining the sum insured. The style of the building and materials to be used in any reinstatement may have very special cost implications. If the elevations have special qualities they may have to be preserved after a fire at considerable expense. Total demolition may not be permitted. Reinstatement may extend to restoration including salvaging and conserving ornamental timber or plaster works. The valuation should reflect all risks so that the sum insured is adequate in extreme circumstances.

Domestic buildings

The sum insured for fire insurance of a domestic house or flat is dealt with by most people at some time in their lives. The principles of assessment are similar to those already stated, but the reasons for a marked discrepancy between purchase price and the proper sum insured should be understood.

The cost of construction of a single detached and isolated house will provide the basis for the calculations as previously described and this cost can be related to purchase price. Generally speaking, however, the construction of a house is as part of an estate or group of buildings. In some instances the development of a very large area is determined and construction of similar types of units is undertaken over a relatively long period. The designer and builder have all the facilities to construct at very competitive prices. Bulk orders for materials can be placed, gangs of workmen can be organised in the most effective way and large items of plant can be used to great advantage. Most suburbs and new towns are developed in this manner and this mass production of units determines the purchase price of the house.

The fire insurance policy for a single house, however, is an individual contract between the insurer and the owner of the property. In the event of damage by fire the owner is faced with employing an architect and builder to deal with one house only, whether it is detached, semidetached or one of a terrace. No longer is the architect concerned with hundreds of houses but just one, and no longer can the builder order materials in bulk and organise large gangs of workmen. The whole process of reconstruction becomes considerably more expensive.

Many domestic insurance policies do not invoke average, but in these situations the policy will probably include a clause allowing the insurers to repudiate the contract unless the sum insured is adequate for total reinstatement.

In dealing with a block of flats a consideration similar to the house situation arises. If the insurance contract is arranged separately flat by flat, very severe cost implications can develop. Fire damage to a penthouse on the top of a 20-storey block of flats could mean the need for high cranes and expensive scaffolding in the reinstatement and all these eventualities would have to be reflected in the sum insured.

It is now common practice for the residents in blocks of flats to form themselves into a company. One of the responsibilities of such a company is to arrange fire insurance for the whole building. This allows for damage to one flat to be regarded as part only of the whole sum insured and the sum insured is related to reinstatement of the complete block and not to one flat. In the situation of a block of eight flats where the sum insured is £320 000, the expenditure in one flat of £60 000 for fire damage repair would be accepted by the insurers, although only £40 000 would appear to be allocated to each flat.

Industrial estates

The industrial estate provides a similar situation to the housing estate. Experience has shown that the adequate sum insured on a separate industrial unit of a large estate should be approximately 50% higher than the proportionate cost of the original construction at the moment that the fire insurance responsibility is transferred from the contractor to the owner or occupier. Apart from the additional costs already mentioned in connection with housing estates, it is particularly important to appreciate that the removal and reinstatement of a limited number of concrete or steel main frames in one industrial unit is a very expensive operation.

It is unusual and impracticable for users of an industrial estate to combine fire insurance as described for the flat complex. The grading of the use or production in each unit would affect the fire risk and the hazard and premiums would have to be considered separately. In a block of flats, however, the use and risk is constant throughout all units. A situation could arise with a terrace of five industrial units each occupied by different tenants. The valuation to be prepared is for the landlord, who insures the whole block with only one sum insured. As long as the use and name of each tenant is disclosed, the insurers will cover the risk as one sum with a whole block policy for the whole terrace. As tenants change, insurers must be notified and, depending on the record and use of the new tenant, so the premium may change.

Fine art

Works that have an aesthetic value of their own demand special consideration by the valuer in recommending the sum insured for material damage. This special consideration would extend to ceiling and wall painting, plasterwork, timber panelling, sculpture, engraving and other finishings and materials which could not be reinstated on an equal value basis. The loss of items or areas regarded as fine art or similar should be considered as an additional and special risk. The particular risk should be valued by an expert in that particular subject and where the insured requires cover beyond reinstatement, then a separate indemnity insurance risk should be arranged.

There are circumstances where the insured would only require reinstatement on the basis of a similar but modern treatment. Wall panelling in timber taken from a famous naval ship could not be reinstated with the same type of timber. The matter should be explained to the insured and if reinstatement is only required in current available timber, then the valuation report should fully describe the basis of the sum insured and disclosed to the insurer. In a perverse way, if this advice is ignored and very rare materials are retained after partial loss from fire elsewhere in a building, then average may apply.

Similarly, the severance of a unique building from the land it stands on by total fire destruction may produce a situation where the combined value of the land and the reconstruction costs would not give an overall value equal to that before the damage. Reinstatement as such may not put the insured back in the position before the fire.

Indemnity based on value may be the financial solution, provided that the sum insured is adequate and complete disclosure of all details took place at the time the policy was taken out.

Consequential loss

In this chapter the subject throughout has been the material damage arising from the effect of fire on buildings. The same principles would apply to plant and machinery. Two other widely known insurances are personal injury and consequential loss. Both of these insurances are outside the scope of the architect and surveyor except in one respect. Consequential loss could be described as the 'knock-on' effect from material or personal damage. A simple example would be the effect of reductions in the profit of a commercial undertaking because the insured was deprived of the use of the building as a result of a fire. Included in those reductions of profit might be the need and cost to provide alternative accommodation while the building was reinstated. Alternatively, the need to reinstate the material might be very urgent in order to ensure that the plant in the building is put back into full operation in the shortest time possible. Additionally, if the building damaged by fire was leased, provision for loss of rent to the landlord may need to be considered; this latter point is often covered under the material loss policy of the building, by providing a sum equivalent to the rent multiplied by the estimated time to rebuild. Calculations for consequential loss can become very involved and complicated, but usually they are related in some way to an estimate of the total length of disturbance to the existing business resulting from a fire. The valuer may therefore be asked to advise on the maximum time necessary to rebuild in the event of fire, so that a view can be taken on the maximum time during which profits would be affected. It is essential that the valuer eliminates all general consequential loss provisions from the valuation report and their effect on building costs.

Where a building is leased, and the landlord pays the insurance premium for material damage (i.e. reinstatement) and the tenant pays the insurance premium for consequential loss, a difficult problem could arise in the event of a fire claim. The insurer for the tenant's consequential loss may wish to contribute substantially to the cost of the reinstatement of the material damage in order to speed the recommencement of the business. The reinstatement would be in the hands of the landlord and the contribution by the tenant's insurers for consequential loss would need very careful negotiation and contractual operation. This kind of eventuality should be clearly defined at the time of placing the risk and not left for diverse opinion following damage by fire.

It is possible to make provision in the policy for loss arising from the interruption of the insured's business by an insured event. If this is required it should be raised with the insurance broker.

A typical business interruption clause in an insurance policy would provide that: if the business is interrupted or interfered with as a direct result of loss, destruction or damage by an insured event for which the insurer has admitted liability in accordance with the perils of policy, then the insurer will pay the amount of loss resulting from the interruption or interference according to the provisions of the policy.

In the action in tort of *Dominion Mosaics Co. Ltd* v. *Trafalgar Trucking Co.* [1990], the building was destroyed by the negligent conduct of Trafalgar, who admitted liability. The premises had been valued at £80 000 prior to the fire and at £20 000 after the fire. Rebuilding costs would have amounted to £570 000 and without alternative premises during the rebuilding period Dominion would have lost £300 000 per annum in profits. Dominion, however, chose immediately to buy new premises at a cost of £390 000 and were able to operate their business without incurring any loss of profits. The Court of Appeal held in favour of Dominion in the total sum of £390 000 being the actual cost they paid out and having incurred no loss of profits. It would appear to be a reasonable assumption that if the policy had not included a provision for consequential loss, then the loss may have been restricted to £60 000 (indemnity based on market values) or the cost of reinstatement if a suitable memorandum was included in the policy, without any compensation for loss of profits. The residual value of the damaged building and land in the sum of £20 000 appears to have been disregarded in the judgment.

It is accepted in law that the loss of the insured is related to and should be measured at the time the loss took place. Subsequent action after the time of the loss in the Dominion case, but before the court trial, showed that Dominion invested about £100 000 in their new building and later sold it for £690 000. On a profit and loss account, this showed an overall profit of £200 000 instead of the loss of £390 000 claimed. The Court of Appeal held that the plaintiff (Dominion) was entitled to the £390 000 and held no obligation to account for conduct after the loss that is not directly related to the property of the action. This may be regarded as an extreme illustration of the outcome of consequential loss, which is a matter to be considered by experts on the precise details of each particular case.

The expert approach

The valuation report should never be taken as a routine duty with standard clauses being included without careful consideration. There are many examples to demonstrate how the expert approach will save the insured premium and also obtain the greatest cover for the risk being insured. The following situation is but one example.

A disused brick-built six-storey warehouse overlooking a river has been granted planning permission for conversion into domestic flats. The planning requires the upper storey to be removed together with the pitched slated roof, which currently leaks and where the supporting timbers are in an advanced state of decay. The foundations, steel framing, brick walls and intermediate timber floors are in a reasonable state but require some repair and adaptation to suit the proposed new use. Among the building items that will be taken out and disposed of in the refurbishment are all internal finishings, ceilings, mechanical and electrical services, sprinklers, hoists, lift, doors and windows. The insured requires insurance cover on a reinstatement basis.

The valuation report must explain that in order to qualify for the reimbursement of the costs of the reinstatement, reinstatement has to take place. Reinstatement will not take place in respect of the upper storey, roof, finishings, services, sprinklers, hoists, lift, doors and windows and in respect of these items the valuation should provide a cost for demolition, debris removal and temporary protection against the elements. The sections of the structure to be retained and repaired would be included in the valuation on a reinstatement basis. The resulting sum insured may be less than 50% of the total reconstruction but it would reflect the true position and no waste in the insurance premium would have taken place in respect of items not qualifying for insurance reinstatement.

The valuation must detail the items at risk that are being insured, those items being excluded and the reasons why, so that application of average is not used in a perverse manner by the insurer. Effectively this is half-way to a first loss policy, but different in that all insurable items of risk are being fully covered.

Prevention costs

The expert approach may extend to further considerations. An important consideration is the cost of preventing damage to insured property once a fire has started. In *Symington* v. *Union Insurance of Canton* (1928), a cargo of cork was to be loaded onto a ship when a fire broke out on the jetty. To prevent the fire from spreading some of the undamaged cork was jettisoned. The policy, a marine policy, covered 'loss or damage to the interest insured which may reasonably be attributed to fire'. The Court held that the fire was an imminent peril and the proximate cause of the damage to the cork and thus the claim fell within the policy wording. There is no clear English authority in nonmarine property insurance cases although, until recently, academic opinion strongly favoured the view that principles would be likely to apply. The extent to which that confidence will survive the Court of Appeal's decision, not yet formally reported, in *Yorkshire Water Services Ltd* v. *Sun Alliance and London Insurance Ltd* and others, is perhaps questionable. In the Yorkshire water case, decided in 1996, the costs of preventing flooding to third party property following the collapse into a river of a sewage sludge tip were held not to be covered under the terms of a public liability policy, although had flooding occurred before expensive avoidance measures had been taken following the collapse, it would seem that the policy would have had to respond to the claim. Clearly a distinction can be drawn between a liability policy and first party property cover, but nevertheless the position remains unclear in relation to such property policies.

The practical importance can be illustrated where there are, for example, two large industrial buildings joined by a small link structure. A practical decision might be taken to destroy the link building to prevent the spread of a fire from one of the main buildings to the other, before the link itself has been affected by the fire. It would seem harsh on the insured to deny cover for the loss arising from such deliberate destruction. To be safe, the point should be raised with insurers, through the brokers, prior to the inception of the policy. If necessary, express provision can be made in the policy although some insurers may require an additional premium. The adequacy of the sum to be insured would also require expert opinion, particularly if the link and one main building were insured separately from the other main building.

Professional valuations

This chapter has shown the general approach required in preparing a valuation report in order to determine the sum insured for buildings. Most prudent undertakings have professional valuations prepared and the advantages can be summarised as follows:

- The problems of a particular building or complex are reviewed in depth by an expert.
- The sum insured is accurate, giving adequate cover without waste of premium from overprovision.
- In the event of fire, the valuation will be generally accepted by the insurer as a sound basis for negotiation without application of average.
- The valuation provides an accurate basis for adjusting the sum insured during ensuing years.
- Measurements and details of construction are recorded in a proper manner and will be of great use for reference in preparing an insurance claim in the event of fire damage.
- Where variations in premium rates apply because of different types of construction or different occupancies of the buildings, the valuation of each building separately provides a basis for calculation of the fire premium.

It is good practice to have professional valuations reviewed annually.

After the initial inspection and valuation the next two or three years can be adjusted by reference to building cost indices and each three or four years adjusted by a site inspection and detailed reconsideration. Regular inspection of large factory complexes ensures that the inevitable changes that take place in buildings are properly adjusted in the sum insured.

In fire insurance valuations, large sums of money are usually involved and the professional adviser is himself at some risk. He should notify his own professional indemnity insurers of the extent of his work in this field and ensure that his insurance cover is adequate.

"Fire Insurance? No, that'll be our other establishment."
brian bagnall

Chapter 10
Disclosure and Placing the Risk

Disclosure in law

A properly completed insurance policy is a contract. The consideration is the premium paid to the insurer who in return undertakes to pay the insured a sum of money by way of indemnity upon damage or loss arising from a specified event; for example, material damage by fire. All legal contracts must observe good faith by both parties in all relevant matters. In respect of contracts of insurance the insurer, who is taking the risk, relies mainly upon the information provided by the insured. That information is therefore considered to be given in the utmost good faith and these material facts are referred to as disclosure.

The law requires disclosure of every circumstance that is material and would influence the judgment of a prudent insurer in fixing the premium or determining whether to take the risk.

Many statements that endeavour to be all embracing become ambiguous by the very nature of attempting to be both precise and far ranging. The word 'influence' in the above requirement is such a case. The insured discloses the circumstances, but the insurer makes the judgment. Does the insured disclose only the circumstances that they consider will influence the judgment by the insurer on a 'required to know' basis? Should the insured disclose an exhaustive list of items in addition to answering standard questions on a proposal form to enable the insurer to be totally advised? Should circumstances be disclosed by the insured to enable the insurer to disregard them, to decide not to take the risk or to charge a higher premium?

It was established in the House of Lords in *Pan Atlantic Insurance Co. Ltd* v. *Pine Top Insurance Co. Ltd* [1995], on the matter of disclosure, that an insurer can avoid liability only if two requirements are satisfied:

- the judgment of a prudent insurer would have been influenced by the information not disclosed; and
- the insurer in question was in fact induced by the insured's presentation of the risk to enter into the insurance contract.

Disclosure and the building

Full and correct disclosure is therefore vital at the time of placing the risk if a subsequent claim is to be maintained and for payment to be received from insurers. Disclosure in respect of insuring material damage to buildings would cover the following:

- age of building – general condition relating to good or poor repair;
- vacant, unoccupied – maintenance of security, detection and other alarms;
- fully or partially occupied;
- use – process, materials used;
- materials employed in production;
- description of dangerous materials stored – explosives, petroleum and gas products, plastics;
- change of tenant;
- residential properties other than owner/occupier – used by students, DHSS purposes, holiday lets;
- multi-tenure structures – ground floor shops with upper storey domestic, terraced or semidetached industrial units, commercial buildings;
- construction – walls, frame, roof, floors and structural components; large timber elements of construction or finish; large areas of glass, shop fronts, curtain wall glazing, atrium construction;
- heating arrangements – air conditioning, low pressure radiators, gas fires, electric fires, portable fires, open fire places;

- maintenance – specific requirements scheduled on the fire certificate; regular inspection of hand appliances, detection and warning systems; general maintenance of fire exits, lighting and direction signs; regular drill for emergency evacuation;
- quality and pressure of water supplies to owner's fire fighting equipment – the grade of sprinkler installation;
- situation of building – urban or country, distance to the nearest fire brigade;
- record of past insurance claims;
- existence of a compulsory purchase order on the building;
- published future planning decisions affecting the property;
- listed building – conservation orders;
- VAT status of insured;
- a copy of a valid fire certificate or amended fire certificate.

The insured must not restrict the disclosure to answering the questions on a proposal form on the basis that if a relevant question is not asked then that information is not required. All matters that a prudent insurer would wish to know must be disclosed in a full and honest manner. If special conditions or explanations are required, ensure that they are clearly set down in writing and exchanged by the parties. Should the insurer ask questions that are not applicable, then the insured should state why they are not applicable.

Any change to the information provided at the commencement of the policy cover must be disclosed to the insurer when that change occurs. On renewal, all information should be checked and it should be ensured that all changes have been notified. The insured should always have an annual meeting with the broker and review in depth all insurance requirements, matters and sums insured.

Insurance, construction and contents

A fireproof building probably does not exist, but a fire-resisting building is closer to reality when viewed as an insurance risk. A fire-resisting construction or massive type construction is a structure composed of noncombustible materials, incorporating:

- protection of iron or steelwork
- protection of floor openings

- protection of exterior openings
- avoidance of wood trim, finish and wood transoms.

A building constructed entirely of fire- and heat-resisting materials is fire-resisting as long as it contains no combustible contents. Many contents are combustible and are liable to destruction by fire. These are the principal risks that insurers assess in determining the premium to be paid. Insurers give good discounts for fire-resisting structures and substandard construction is penalised.

Risk control

Before placing the risk with insurers, the overall situation should be considered by the insured with a view to improving, where possible, the disclosure of the facts. This is risk control and is effectively part of risk management of the buildings on which the insurers will quote a premium. An insurance broker may advise on risk control. Such advice could extend to:

- comparative costs for sprinkler installation to saving on premiums;
- adequate and safe storage of dangerous materials;
- proper protection to buildings that are vacant; sprinklers and plumbing shutdown; electrical supplies switched off at meter and padlocked in place; adequate locks, temporary window shutters to prevent arson;
- maintenance contracts in place for hand and fixed detection, warning and extinguishing appliances and installations;
- regular compliance by the insured with all requirements scheduled on the fire certificate.

Risk control should be in the mind of the insured as a constant obligation, so that the risk of damage from fire to the building is reduced by proper maintenance and other procedures, enabling a good and tidy disclosure to the insurer on a regular basis.

Risk management

Risk control can be extended to the more extensive function of risk management. Consultant specialists can advise on the wider subjects of risk assessment, risk elimination or reduction, financing and funding alternatives to insurance and contingency planning services should existing facilities become unavailable from fire damage. This advice would be rendered in respect of a particular location and is specific to the matter being reported.

Insurance broker

In placing the risk with insurers, the large firms or organisations are likely to arrange the business through an insurance broker. The broker's primary function is to arrange cover and to obtain the best rates and terms available. They will also advise to varying degrees on risk management. When it comes to claims, brokers do have a part to play in negotiating between the policy holder and the insurer. They do not get involved in assembling or valuing the claim, but they will assist in the interpretation of cover and are often in the best position to negotiate difficult issues with the insurers. The broker may have some liability if they have been professionally negligent.

Disclosure and the broker

The broker should assist the insured in providing full disclosure. If the risk produces a high insurance premium, commercially the insurers may be keen to acquire the business and important items of disclosure may not be questioned before accepting the premium. A large claim, however, will produce those very questions and the insured will find that the responsibility for disclosing all relevant facts lies with the insured. A proposal form is the basis of contract in most cases, but the insured's duty of disclosure goes beyond the standard questions asked. All relevant matters must be disclosed.

A simple case of a domestic building not only being used as the family dwelling but also as an office may be sufficient for insurers to resist and avoid a claim. A further example could be a part time kindergarten or nursery school at a dwelling. If, however, at the time of placing the risk, disclosure that a room was from time to time used as an office or kindergarten, the premium would have probably remained unchanged and the claim would be valid.

The situation can however, become more serious when insured have to answer to regulations and legislation. Again, using a dwelling as an example, should insured receive rent for a furnished country cottage or seaside apartment, there is an obligation to maintain legal standards both in respect of the buildings and some of the contents. Compliance with these obligations will need to be disclosed to insurers. Should these obligations be avoided, not only would the insured be liable for legal penalty, but insurers are most likely to reject any claim arising from fire damage. Because of the structure of insurance policies for standard domestic properties, the cover could extend to public liability claims and should there be lack of proper disclosure the insured may become personally liable for large financial liabilities.

The broker will no doubt ask insured many questions, but should there be lack of full disclosure the insured may find that insurers will avoid a claim.

Non-invalidation clause

The need to disclose matters continues throughout the term of an insurance policy. It is no defence to say that the disclosure at the time the policy commenced was complete and correct if relevant matters change. Brokers should ensure that the policy includes a non-invalidation clause in the following manner:

> 'This insurance shall not be invalidated by any act or omission or by any alteration whereby the risk of damage is increased, unknown to, or beyond the control of the insured, providing that the insured, immediately they become aware thereof, shall give notice to the insurer and pay an additional premium if required.'

'Smoke damage'

Chapter 11
After the Fire

The professionals involved

When a building has suffered damage by fire, certain clauses of the insurance policy may begin to operate. If the damage is very limited and simple, the matter may be resolved directly between the insured and the insurers. However, in many cases of fire the eventual outcome cannot be settled without involving professional advisors to act for both parties. In order to perform in a proper and professional manner it is essential to understand the scope and responsibility of those involved.

The first responsibility must be with the insured, who has to notify the insurer that a claim arises – probably first by telephone and later confirmed by completing a claim form. It is of paramount importance to appreciate the fundamental difference between the objectives of the insurer and those of the insured. The insurer will be dealing solely with financial compensation arising under the terms of the policy; the insured will not only be seeking that compensation but in many cases will also be seeking to reinstate the building from the funds obtained.

The insurer has a commercial responsibility to keep the amount of compensation to the minimum commensurate with his legal obligation and business probity. A loss-adjusting firm will almost certainly be appointed to act and, if the claim hinges on a legal matter, a solicitor as well. The solicitor will need to act in close liaison with the appointed loss adjuster. These professional parties will be appointed directly by the insurer.

On the other hand, the insured is unlikely to be experienced in claims or in reinstatement after fire and will often not know to whom to turn for proper advice or the co-ordination of his requirements. Without a co-ordinator or team leader to direct activities, the insured can be poorly represented and at a disadvantage against a highly experienced professional team acting for the insurer. The architect and quantity surveyor have the qualifications to handle this position, but either or both are often appointed too late or with limited instructions. In addition, insurance matters are on the periphery of their expertise and therefore they tend to await instructions from others instead of taking charge. In fact they should be discharging a duty befitting their professional capabilities.

Chapter 12 discusses the claim, but before dealing with this matter the duties of those involved after the fire should be reviewed. In broad terms they can be summarised as below. The insurer (dealing with a financial claim) may appoint:

- loss adjuster – architect/engineer; quantity surveyor; accountant
- solicitor.

The insured (dealing with reinstatement and compensation or a financial claim) may appoint or need to liaise with:

- insurance broker
- assessor
- architect/engineer
- quantity surveyor
- accountant
- banker
- landlord
- solicitor.

The insurer

The insurer is the person taking the financial risk of damage to the material asset for which a fire insurance premium is paid by the insured. On notification of a claim the insurer will ensure that:

- the policy is currently in force and all premiums due have been paid;

- the details have been issued, clearly defining the conditions of the policy;
- the damage applies to the property insured;
- liability for the damage is covered by the policy;
- third parties (e.g. landlords) named in the policy are notified;
- a loss-adjusting firm is appointed and is provided with all the details of the policy.

The insurer will deal with recommendations from the loss adjuster including action relating to interim and final decisions and payments. If liability for the claim is denied or limited, the insurer should inform the insured as quickly as possible.

The loss adjuster

The loss adjuster is qualified by professional examination through the Chartered Institute of Loss Adjusters. The appointment will be by the insurer and the primary duty is to approve or adjust a financial claim submitted by the insured so that recommendation of payment can be submitted to the insurer.

It is usual for a loss adjuster to receive instructions only from insurers and the majority of loss adjusters have no other type of client. Much of the work would be of a general nature, but larger firms employ adjusters who specialise in particular subjects. A loss adjuster qualified in accountancy would be well suited to deal with consequential loss claims relating to reduction of profits. Likewise, a quantity surveyor working as a loss adjuster would be best suited to dealing with a large fire insurance claim on a building, while someone qualified in law could best handle professional indemnity claims. Many adjusting firms do not employ professionally qualified people to handle all types of loss and therefore rely on professional consultants to assist them from time to time. In these situations the independent professional may receive instructions and payment of fees from the adjuster. However, where the loss is significant or of a very specialised nature, the appointment of the consultant may be directly by the insurer; in these circumstances the consultant would need to work in close co-operation with the loss adjuster.

Because the adjuster's prime object is to limit the liability of his client in accordance with the details of the insurance policy, the duties can be far ranging. At times it may appear that the adjuster assumes responsibilities beyond the professional duties to the client (the insurer), but in fire damage situations the experience of the adjuster is used with the intention of indemnifying the insured so that the material damage is put back into the position equal to that before the damage. The loss adjuster will therefore be concerned with the following duties when called to an incident of fire damage to a building:

- Investigate the cause of the fire and confirm that the insurer is liable and that the policy should come into operation.
- Obtain and study the fire brigade report of the incident and consider the influence it may have on the operation of the policy.
- Ensure that photographs are taken of the damage immediately after the fire.
- Ensure that the maximum is obtained from salvage. This applies particularly to plant and machinery, which can have high resale value either in this country or overseas.
- Approve the method and cost of first stage demolition, temporary protection and support of buildings owned by the insured and third parties.
- Establish whether the claim will be on an indemnity or reinstatement basis.
- Approve the extent of second stage demolition. In this situation an independent architect or engineer may be instructed by the adjuster to give opinions on the extent that structural work needs to be removed and renewed.
- Approve the specification of the work to be carried out to reinstate the building following the fire damage. In certain situations, as described in previous chapters, this could involve the construction of a new building comprising different types of materials, different floor and storey heights and even constructed at a different location.
- Approve the method of obtaining tenders or determining the building costs. The advice of an independent quantity surveyor may be of considerable value on this matter.
- Offer the facilities of discount arrangements on bulk items. Many adjusters have this arrangement with suppliers of carpets, office equipment and similar items.

- Consider the effect of the reinstatement programme on the consequential loss claim and ensure that the overall solution limits the overall liability of the insurer to the minimum. The services of an independent accountant could be essential in advising on the correct balance between obtaining reinstatement costs through competitive tenders based on the most economic reconstruction time, against the loss of production and profits from the business.
- Consider the requests and need for providing alternative accommodation for the insured so that business can continue with the minimum of disturbance. In the case of domestic property a recommendation may be required on the alternative costs of renting temporary accommodation or reimbursing hotel expenses.
- Consider the adequacy of the sum insured and the possibility and effect of applying average.
- Consider excess provisions in the policy.
- Consider and recommend interim payments.
- Correlate the claim for building damage with other insurance claims arising from the same incident and liaise with any other appointed loss adjusters.
- Consider and recommend the financial settlement.

As a result of extensive damage by fire, several policies may come into operation, often with different insurers. It is possible, therefore, and not uncommon, for a fire to invoke separate policies from different insurers on the buildings, plant, machinery, contents and the consequential loss of profit in a business, and for each insurer to appoint a different loss adjuster. There are rules of etiquette in this matter and the insured is not usually placed in a position of disadvantage. However, there can be advantages to the insured in placing his insurance on building, contents and consequential loss with one insurer, so that in the event of a claim for fire, one loss adjuster would be appointed.

The need to understand the business activities of some larger industrial firms has persuaded insurers to appoint loss adjusters on a long-term basis. This ensures that when a loss occurs, the adjuster will be on site within hours and dealing with persons well known to him over a period of years. In addition, the adjuster will understand the working of the business and be in a position to assist in approving proposals to minimise the loss.

The insured

The insured is a client of the insurance industry. The insured pays the insurance premiums and is relieved of the risk. If fire damage is of a strictly limited nature, it may be rectified in a simple manner and both parties may agree that competitive prices from two or three builders will be the basis of the reinstatement and the claim; an example would be smoke damage to a room with redecoration providing the reinstatement. Once the damage is no longer simple, the insured will be wise to employ consultants. In all cases the insured or the consultants should ensure that:

- Proper written notification of the incident is made to the insurers.
- Notification and details of the appointment of other consultants is given to the insurers.
- Where the insured is a tenant, the landlord is informed.
- The full extent of the damage is assessed and the possible effect on the running of a business is understood. This should lead the insured to appointing the consultants best suited to limit the loss, and to reinstate the building and the business as quickly as possible and at the most economical cost. Consultation with the appointed loss adjuster is essential on these matters.
- Prompt attention is given to finding suitable alternative accommodation for business or domestic use where the damage warrants this requirement.

Insurance broker

The principal duty of an insurance broker is to place the risks of the insured with the insurers at the most favourable terms. The details affecting their duties are described in Chapter 10.

Where a claim on a policy arises, the broker will ensure that the interests of the client are protected. Most brokers have a claims department and, depending on the size and nature of the business, have staff who specialise on particular types of claim. This service can be very successful, especially where the claim has an insurance basis in essence without the need for specialist knowledge in the subject of the loss. Where the loss is due to the effects of fire, providing serious and extensive material damage, then the broker will usually recommend the appointment of independent professional firms to handle the claim and the reinstatement.

The insurance broker will need to consider the type of consultant best suited to deal with the insured's case. If the material fire damage is to be settled on an indemnity basis without reinstatement, an assessor or valuation surveyor would be considered. Should reinstatement be the recommended course, then independent building consultants should be considered. The nature of the reinstatement will determine these recommendations, but the lead consultant or project manager should have a good working knowledge of insurance risks, claims and solutions.

Assessor

The assessor could be regarded as the direct counterpart to the loss adjuster, in that both would have responsibility for determining financial matters on insurance but usually neither is directly involved with reinstatement. Where the claim is to be determined by indemnity based on depreciated value of materials, the assessor will schedule the items and values of the damage and present the claim on behalf of the insured to the loss adjuster. The assessor has a good understanding of insurance and the claims that can arise. In addition, some specialise in a particular subject, such as building, and therefore have good knowledge of the construction industry. Their negotiations with loss adjusters are effective, but in the majority of cases their duties are restricted to negotiating the financial compensation, while the responsibility of reinstatement remains with the insured.

Architect and engineer

Where a building is to be repaired and reinstated after fire damage, the insured will appoint an architect or engineer to take the responsibility for protecting the building and arranging for reconstruction. It is essential that the architect's or engineer's brief is clearly defined at an early stage. In particular, there should be a complete understanding of the financial implications of the likely compensation from the policy and the cost of reinstatement or reconstruction. From the beginning the consultant should liaise closely with the loss adjuster and reach written agreement step for step. The duties therefore are likely to be to:

- determine the extent of first stage demolition of any dangerous structure;
- arrange first-aid repairs and protection of the undamaged sections of the building;
- arrange temporary support of adjoining buildings;
- ensure the maximum is obtained from salvage;
- have extensive photographs taken at each stage of demolition, protection and temporary support, and obtain the written agreement of the loss adjuster that these represent a true record of the damage and extent of work required for rectification;
- determine the brief with the client and, assuming reinstatement is to take place, proceed with the following:
  - arrange for second stage demolition or total site clearance;
  - prepare specifications and drawings for the reinstatement and obtain all necessary consents and approvals;
  - determine with the loss adjuster the basis and programme of the building contract and obtain tenders or negotiate a contract;
  - supervise the reinstatement.

Quantity surveyor

The duties of the quantity surveyor will be determined by whether the material damage to the building is to be reinstated. If reinstatement is not to take place, the quantity surveyor will have duties identical to those of an assessor. A financial claim will need to be prepared for negotiation and agreement with the loss adjuster.

If reinstatement is to take place, the quantity surveyor will act in the normal way, in conjunction with the architect, in preparing documents to obtain tenders and will subsequently assist with the running of the contract. The cost evidence and information from the building reinstatement contract may be sufficient to prepare the insurance claim and get it agreed with the loss adjuster. If this is not the case then a separate document may be required defining the extent and cost of the damage recoverable under the policy.

The quantity surveyor should consider the following:

- the financial implications of insurance policy and correct understanding of the clauses, confirming whether the policy is based on indemnity or reinstatement;
- the effect of average;
- the effect of deductible excess;
- the need to prepare a priced schedule of loss;
- the programme of reinstatement work relating to any consequential loss situation;
- the effect of VAT on builders' accounts and whether it is recoverable under the wording of the policy;
- the correct application of policy requirements to recover properly inflationary costs, loss of rent and cost of alternative accommodation;
- the preparation and agreement with the loss adjuster of the procurement document, to obtain tenders or to negotiate a tender for the reinstatement work and the subsequent need to prepare contract documents, interim valuations and final account;
- the whole financial programme, agreeing it step-by-step in writing with the loss adjuster before committing the client to expense – in the case of reinstatement, the proper arrangements for interim payments from insurers;
- the need to define carefully the duties of each consultant and establish who will be responsible for paying respective fees where more than one party is insured or where there is more than one insurer.

Advisors and interested parties

Although the insured is primarily responsible for initiating the claim and reinstatement, it must be accepted that, having suffered the loss of the business or home, actions may not be rational. Therefore whoever takes charge in the first day or two should keep an open and clear mind regarding the necessary actions to be taken. The period immediately after the fire is crucial.

Landlords, bankers and mortgagers must all be advised and their interests disclosed. If the damage is to a manufacturing business then customers and suppliers must be advised. Loss of contents and stock will need to be scheduled and probable consequential damage to business profits will need to be reviewed. In a complicated incident the services of an accountant will be essential. His duty will be to examine the financial effect of each course of action recommended so that a rational overall solution is determined. Architects and quantity surveyors tend to consider building costs and programme as paramount over all other factors. In many instances the financial loss due to waste of stock or decline in business can be a decisive factor in determining the correct and most economic remedy.

The solicitor

Where a dispute on liability or interpretation of the policy cannot be settled between the insured and the insurer or their representatives, the matter may then develop into a legal action. Compared to the vast number of fire insurance claims, these circumstances are unusual and even at this stage there is a preference on the part of insurers to settle out of Court.

Quite apart from the enormous costs and delay of litigation, it has to be recognised that case law on insurance issues (as in many other areas of the law) is extensive and individual decisions are often simply a reflection of the facts of the particular case without necessarily establishing principles that can be readily applied to other circumstances or policy wordings. Insurance law is also subject, with certain exceptions and many refinements, to the general law of contract that has been developing for centuries through many thousands of cases and statutory interventions. One major consequence of the complex way in which insurance law has developed is that any litigation will carry a significant element of uncertainty. One should go into litigation with eyes wide open to these difficulties.

In cases of dispute on fire damage related to design, most construction professionals have insurance policies for professional indemnity. Many are required through membership of their professional institutions or societies to take out professional indemnity insurance policies with a stipulated minimum sum insured. These policies invariably require the insured to notify the insurer as soon as a claim or dispute arises. If the matter proceeds, the insurer will do their utmost to reach a commercial solution without reference to the courts.

Action through law can often be protracted and may concern both liability and quantum. The responsibility for the actual reinstatement (if at issue) will always rest with the insured, and it should be appreciated that recourse to lawyers can often delay the financial settlement of an insurance claim and therefore deny the insured the advantage of availability of cash to rebuild. Where the matter of a claim for fire damage to a building is put to lawyers to resolve, parties should bear in mind the following:

- The solicitors appointed should be knowledgeable and experienced in insurance and building matters.
- Commercial negotiations may still continue independent of the legal action.
- The parties to the dispute should inform the solicitors of the circumstances in clear terms so that the lawyers can determine the points at issue. Litigants should avoid attempting to decide on legal issues.
- Solicitors should determine the best practical approach to protect the client's interest. This usually means, in the case of the plaintiff, commencing proceedings as quickly as possible so that the points at issue are properly defined, thus crystallising the parties' attitudes towards settlement. The defendant (inevitably the insurer) will equally want the matter defined in legal terms at the earliest opportunity.
- Considerable legal costs in these cases are incurred before trial. In broad terms, a successful litigant will not recover all legal costs and the reimbursement may be as low as 50–60% of the total.
- Contributions to or from third parties must be clearly defined.

Professional fees

Professional fees charged for preparing, submitting and negotiating a claim for financial compensation are not recoverable by the insured through a normal insurance policy. Professional fees directly associated with the reinstatement of the building after fire damage, however, are recoverable by the insured through a normal insurance policy. Where a solicitor is appointed and the case proceeds to a court hearing, the costs of the case and the judgment or award will be at the direction of the judge or arbitrator. These costs would therefore be settled outside the scope of the normal insurance policy.

Insurance, contract and tort

Uppark was built in its commanding position high on the South Downs in 1690 for the Earl of Tankerville. Subsequent owners made few substantial changes to the house, which at all times was kept in good repair. In 1954 Uppark was given to the National Trust.

Disaster struck Uppark in August 1989, when it was severely damaged by fire. Almost all the contents from the ground floor and basement were saved, but the greatest tragedy was the loss of the private possessions from the upper floors which, despite the best efforts of the fire fighters, could not be saved due to the speed with which the fire spread through the house.

As so much was saved of the original Uppark, the decision was taken to restore it. The repair has been the most complicated ever undertaken by the National Trust and was entirely funded by insurance and litigation. The aim has been to return Uppark, in so far as was practical, to its state before the fire. Old and new have been carefully matched so that the rooms may be seen as a whole, creating a suitable setting for the fine furniture that now resides back in its intended position.

The restoration was entirely authentic and the cost was paid for in full by insurance, estimated at about £20m.

Apart from demonstrating the degree of the extent of a reinstatement policy, in order to indemnify the insured so that they were put in the position before the material damage occurred, the legal claim makes interesting reading.

The case is *National Trust* v. *Haden Young Ltd* (26 July 1994), Court of Appeal. The fire was started by the negligent use of a workman's blowtorch, on the penultimate day of a six-month contract to repair the lead on the roof. The contract used in this case was JCT Minor Works. Haden Young was a subcontractor and had been working on the roof when the workmen had gone for a tea break without taking appropriate precautions. The National Trust sued Haden Young for negligence. Haden Young admitted negligence but contended that the terms of the main contract relieved the main contractor and therefore themselves from liability. In the alternative, they argued that they had a contract with the National Trust which had the same effect.

Clause 6.2 of the main contract provided that the contractor should both be liable for and indemnify the National Trust against damage to any property caused by its default. Clause 6.3B required the National Trust to insure existing structures against loss or damage caused by fire, amongst other risks. Haden Young argued that reading the two clauses together the contractor was not liable under Clause 6.2 where the cause of the loss was an event within the scope of Clause 6.3B. The Court of Appeal rejected this argument and held Haden Young liable.

Example negotiation

A shop has been vacated by the tenant with a lease to expire in 15 years. The building is insured with a reinstatement policy in the name of the tenant and is severely damaged, requiring reinstatement to 75% of the building. The tenant wishes to surrender the lease with a reverse premium payment. The landlord will accept a reasonable reverse premium of £190 000, or will require reinstatement, which will cost £250 000.

In this example situation, the insurers may be prepared to accept an indemnity settlement at the lower figure, bearing in mind that if it is rejected the insured has no alternative other than to reinstate and claim £250 000 from the insurers. Provided the sum insured is adequate, the insurers are likely to look at the negotiation in a commercial light and settle for indemnity at the lower figure of £190 000.

'Credibility of information'

Chapter 12

The Claim and the Reinstatement

Principle of indemnity

Everyone concerned with a fire claim should remember, as stated in Chapter 8, that with rare exceptions all property insurance is based on the broad principle of indemnity. Within this context, claims are met either by a cash settlement unrelated to the actual use of that cash, or by reference to reinstatement costs incurred. While settlement on a reinstatement basis is a means of indemnifying the insured, in the discussions which follow in this chapter the expressions 'indemnity' and 'reinstatement' are generally used to distinguish between these two common types of claim settlement.

It should also be remembered that the total sum assured places an overall limit on the level of indemnity recoverable.

At the beginning of this book and in Chapter 11, the reader is alerted to the dangers of trying to extract legal principles from individual cases, which may depend on their own facts and policy wordings. Consideration of the illustrative cases mentioned in this chapter demonstrates the flexibility of the law, perhaps more clearly than elsewhere in this book.

Meet the loss adjuster

The insured, through his professional consultants, would be advised to meet the loss adjuster immediately after the fire brigade has dealt with the site situation and the building or remains have been made safe to the public. Experience will show that a prepared agenda will greatly assist this first meeting. An example of the items to be resolved on a reinstatement claim would be:

(1) terms of insurance policy:
- adequacy of the sum insured;
- excess;
- inflation;

(2) basis of claim (to be reinstatement as against indemnity);

(3) scope of works:
- initial emergency work;
- demolition and debris removal;
- protection;
- shoring;
- salvage;
- full reinstatement;

(4) programme, taking account of consequential loss claims;

(5) methods of procurement:
- negotiation;
- competition;

(6) contractual arrangements;

(7) presentation of costs;

(8) current building regulations;

(9) interim payments to insured;

(10) professional fees:
- architect;
- engineer (structural and services);
- planning supervisor;
- quantity surveyor;

(11) VAT status of all parties;

(12) AOB.

The purpose of a prepared agenda is to concentrate the mind so that the whole programme for the reinstatement and payment of the costs is properly discussed, agreed and then recorded in minutes or by letter.

A situation may arise on any of the above points that overshadows the extent of the fire damage. The fabric of an industrial building may be extensively damaged, whereas the plant it houses is still serviceable at little cost. A temporary scaffold-constructed roof might provide the solution to ensuring that the plant continues to operate. The cost of this temporary roof and wall protection may in itself cost as much or more than the material reinstatement. No doubt this is a case for a claim against the consequential loss policy, but it would have to be agreed from the onset. This type of situation becomes more complicated when the material damage policy is with one insurer and the consequential loss policy is with another. Two firms of loss adjusters would probably be employed, resulting in even greater need for the consultants to the insured to obtain written confirmation on agreements to reimburse all costs and to identify which insurer will pay for each element of the temporary works and the permanent reinstatement.

Basis of claim

The responsibility for preparing and presenting the financial claim under any insurance policy rests entirely with the insured. The need to determine the basis of the claim is paramount at an early stage. The alternatives are:

- reinstatement costs
- indemnity cash payment.

Reinstatement costs

Where the insurance policy contains a reinstatement memorandum, the insured has the opportunity of recouping the cost of putting back the building with similar materials and in a similar style to that which existed before fire damage – in simple terms, 'new for old'. This will apply to all aspects of the building, including items having a limited life such as decorations and service installations.

Reinstatement has a special meaning when used in insurance. In literal terms, reinstatement means to re-establish in former position and to restore in proper order. Insurers usually consider reinstatement in this strict manner. Architects and quantity surveyors often confuse reinstatement with reconstruction. Where the reconstruction ensures that the style, size and performance are similar to those before fire damage, the insurers will regard this as reinstatement. However, where damage is extensive, there is the opportunity for reconstructing in a different way. Perhaps reinforced concrete may be used *in lieu* of timber construction or the size of the building may be altered. Then reconstruction is not pure reinstatement, but contains some degree of betterment, and in this situation a financial reduction may have to be made from the reconstruction costs. Where these alterations are made to comply with building regulations or planning requirements they are accepted as reinstatement, providing the policy contains the 'local authority' clause. But if the changes are a choice of the insured, then the whole reconstruction is not regarded as reinstatement.

Once reinstatement is established and agreed by the loss adjuster, the documentation follows the course of normal building work. The designer prepares drawings and specification notes describing the extent of the work and these are agreed with the loss adjuster. The quantity surveyor prepares bills of quantities, or other suitable documents for competitive or negotiated tender, and the interim certification proceeds as normal.

The contract to reinstate will be between the insured and the building contractor. Likewise the engagement of professional consultants will lie directly with the insured. Payment by the insurers is regulated by expenditure by the insured. This applies to fees as well as to building costs. It is usual to arrange for the loss adjuster to recommend interim payments from the insurer that will follow closely on architect's certificates. These payments will normally be from the insurer to the insured, but there may be cases when the insurer will decide to pay the building contractor and professional consultants direct. The total expenditure on the work will form the claim and in such a case the insured should be reimbursed his total outlay.

Where a building is reinstated it is unusual for payment to be made by the insurer in advance of work being carried out. However, there is a point of view that, as the policies provide for an indemnity, the full indemnity, when determined, should be paid in advance of reinstatement. The balance arising from the reinstatement memorandum over the indemnity value then becomes due if the cost of reinstatement exceeds the indemnity. Although many adjusters resist this view, it is a matter for negotiation and underlines the real necessity to agree fact and procedure with other professionals at an early stage and in a regular way.

It is worthy of repetition to state that reinstatement can be accepted on an alternative site and in a modern design if planning requirements and building regulations are to be properly met. Once the principle has been agreed with the loss adjuster, the procedure will be similar to that already outlined.

Indemnity cash payment

The indemnity cash settlement provides for a lump sum payment as compensation under the terms of the policy. Once the indemnity sum is agreed it is paid in full by the insurer and his obligation is fulfilled. How the money is used and what happens to the damaged building is no longer any concern of the insurer. The basis of this compensation can be derived from:

- a priced schedule of loss
- reconstruction costs
- market values.

Priced schedule of loss

A priced schedule of loss represents the notional reinstatement cost of fire damage to a building. The schedule can be prepared as a specification, as a bill of quantities or just as a schedule of items. It will contain all the items necessary to reinstate the building exactly as it existed before the damage. Whether the materials described would be used in the reconstruction is not relevant. The schedule might therefore contain items of lead plumbing, king post roof trusses, hardwood beams and joinery, timber lath and plaster, and many other items no longer used in modern day construction. The information for the schedule would have to be drawn from the items of debris, past drawings and photographs, if available, and from the question and answer with the insured.

The document will almost certainly be unsuitable for obtaining tenders and will need to be priced in a notional manner. The total cost of the schedule will represent the gross reinstatement cost of the damage using similar materials and in a similar style to existing. If the policy is one of indemnity, the pricing will be related to the date of fire. If the policy contains a reinstatement memorandum, the basis of pricing will include allowance for inflation during the period of planning and reinstatement.

Once this document has been checked and agreed by the loss adjuster, the question of depreciation (or betterment or dilapidation) will need to be considered. This is the point at which the reinstatement cost is converted into indemnity value. Loss adjusters generally have a very realistic view of this matter. Bricks, stone, concrete, slate, steel, glass and many other materials used in a building seldom lose their qualities unless the whole building is allowed to decay. Electric wiring, heating installations and decorations, however, depreciate with age and it is usually to these items that a good negotiator can restrict the adjuster on dilapidations. The adjuster may consider each item for depreciation, but it is usual for negotiations to centre around a percentage figure to be applied to the whole, or sections of, the priced schedule.

It is not possible to obtain a cash payment from an insurer equivalent to the reinstatement cost if the reinstatement is not carried out, even though the policy may contain all the right clauses. The cash payment will always be reduced to the indemnity basis of valuation.

Various legal cases are worth mentioning in this respect, namely *Reynolds and Another* v. *Phoenix Assurance Co. Ltd* [1978] and *Pleasurama Ltd* v. *Sun Alliance Insurance Co. Ltd* [1979]. These cases are mentioned because unsuccessful attempts have been made by insurers to set aside indemnity cash payments related to notional reinstatement costs depreciated for either wear and tear or other factors.

In the Reynolds case, the judge decided that the policy did not give an automatic entitlement to recover the costs of reinstatement. In the particular circumstances, however, where the plaintiffs generally intended to reinstate the property, a true indemnity was the cost of such reinstatement. In the Pleasurama action, a similar decision was reached on the basis that the insured wanted to use the property in the same form as prior to the loss. These cases contrast with *Exchange Theatre Ltd* v. *Iron Trades Mutual Insurance Co. Ltd*

(1983), mentioned in Chapter 8, and the Leppard case referred to below.

Reconstruction costs

In order for reconstruction costs to be used in support of an indemnity cash payment a flexible and understanding approach by the loss adjuster is required. There can be many instances of this type of negotiation and two brief examples can illustrate the possible procedure:

- Following fire damage, the insured decides to reconstruct to provide a larger building while retaining a style similar to existing. By relating floor areas, the reconstruction costs of the whole may provide the reinstatement cost for the section damaged. Depreciation would then apply as before.
- Fire destroys two bays of a complicated industrial building and the insured decides to reconstruct in a simple manner. The reconstruction costs, together with a priced schedule of loss for the complicated sections not reinstated, may then provide the necessary information to agree an indemnity payment for the loss.

Very careful thought should be given to the best method of resolving the claim by way of an indemnity cash payment and the credibility of the information provided is important. Reconstruction costs properly obtained in competition provide a sound basis for use in negotiations.

Market values

If a loss can be related to market values, indemnity may prevail and an option for reinstatement may be discounted. A legal base for this point of view is provided in the case of *Leppard* v. *Excess Insurance Co. Ltd* [1979]. The case shows a cottage totally destroyed by fire and insured in the sum of £14 000. Under the policy, the insurer agreed to provide insurance and indemnity and 'at [their] option by payment reinstatement and repair indemnify the insured in respect of a loss or damage caused by . . . fire'. The cottage had not been reinstated. The cost of reinstatement after taking betterment into account was estimated at £8694. There was admitted evidence that the insured had tried to sell the cottage for a considerable period before the fire at a figure of £4500. The land was valued at £1500. All the figures were agreed by the parties prior to the hearing. The Appeal Court ruled that the correct damage was £3,000, being the difference between the market value before fire (£4500) and the market value of the land (£1500).

Insurers generally respect the adage 'an Englishman's home is his castle'. This relates not only to domestic but also to industrial and commercial properties. If a policy contains a reinstatement provision, in most instances the insurer will agree to reinstatement and refund the costs. Should there be evidence that the insured has previously sought to sell his property, or it can be demonstrated that similar is available without detriment to him, the insurer may look very closely at this situation and consider market values.

Partial loss

The majority of fire damage to buildings represents a partial loss. The principles of reinstatement and indemnity similarly apply. Situations can show advantages by using both. An illustration of this would be an old-fashioned building now used as a printing works, where a serious fire destroys the first floor and roof but leaves the ground floor in a repairable state. Agreement can be reached with the loss adjuster for the ground floor to be repaired on a reinstatement basis, including provision of a suitable roof in lieu of the first-floor construction. Consequently, the upper floor and roof might not be rebuilt as the owner can find better use for the cash compensation. The repair cost of the ground floor storey would be reimbursed in full on a reinstatement basis and the notional reinstatement cost of the first floor storey, reduced for dilapidations, would form an indemnity cash settlement.

Professional fees

Professional fees incurred in connection with reinstatement are usually reclaimable from insurers, as long as the sum insured is adequate and the policy contains provisions for payment of fees.

Professional fees incurred for preparing the claim are not reclaimable through an insurance policy. The cost of preparing a priced schedule of loss would not be admitted in a claim. Indemnity cash settlements do not generally include allowance for professional fees. However, examples have been given in this chapter to show how reconstruction costs can be used as a basis for indemnity settlement, and in these cases the adjuster could recommend reimbursement of fees.

Average

A claim will be subject to the adequacy of the sum insured. Average is explained in Chapter 8, but it is in the claim situation that it is applied in respect of both reinstatement and indemnity.

In the past, some fire insurance policies for commercial and industrial buildings contained a concession on the application of average. This concession provides that if the sum insured represents at least 85% of the total reinstatement cost of the whole, then average will not apply. In these circumstances, the insured was at risk for 15% of the reinstatement costs, but only the last 15% of the total reinstatement. In a complete destruction situation there would have been underinsurance, but it is exceptional for destruction to be 100% complete.

This concession, called the 85% rule, is no longer included in policies, but the concession is still often conceded by insurers, by not applying average where there is partial loss. This concession would have to be sought from and agreed with insurers. However, if such a situation was not conceded, the insurers would need to demonstrate beyond reasonable doubt that the sum insured was inadequate and this would be a matter of opinion that may be difficult to prove one way or the other, bearing in mind that the concession is only 15% of the total.

Excess

Similarly, excess is taken into account when finalising a claim. Generally this is a simple deduction following the agreement of the details of the claim. Complications can arise if the deductible excess relates to 'each and every incident'. In some situations, particularly in cases of malicious damage, like arson, there can be several 'seats of fire' and insurers may take the view that each 'seat of fire' is a separate incident. Circumstances will determine how insurers view such a case and usually they are not severe in their approach. However, bad advice can be given by professionals if all the facts are not properly assembled and fully understood.

Long-term liability

Redundant parts of a building may be damaged by fire but require no immediate attention in repair. An example would be damage to an original plastered ceiling now hidden by a suspended false ceiling. Where the building is leased, the original ceiling may be regarded as part of the structure and insured by the landlord; the suspended ceiling could be the responsibility of, and insured by, the tenant. At the end of the lease, the suspended ceiling may be removed and the damaged plaster ceiling would then be exposed. The landlord would need to protect himself against this deferred situation: this is usually achieved by exchange of correspondence and provision by the insurers of a sum set aside for the defined period. The sum would need to be properly invested to provide an allowance for inflation in building costs.

Fire during construction

If fire occurs during new construction or refurbishment, the claim for loss will be the responsibility of the insured in the normal manner. If the terms of the building contract require the contractor to insure, he will make the claim, whereas the employer will make the claim if he is the insured.

In many cases the building contract requires the insurance for fire damage to be taken out in the joint names of the employer and the contractor. Furthermore, the building contract will stipulate how the money will be administered in the event of a claim being made and paid. Where the contractor decides to use an existing block policy to comply with this condition in the building contract, the name of the employer and the interest they have in the matter must be notified to the insurer. Where the reverse applies and the employer has the responsibility of insuring, the interest of the contractor must be notified to the insurer.

Contractors should be aware of the risks prior to signing a building contract with a central or local government authority where the employer has responsibility for insuring against perils stated in the contract. Many authorities carry the responsibility for insurance themselves, with funds to be provided out of current revenue. Where a substantial loss occurs from fire damage, this being an unexpected liability, the authority may not readily have the funds to implement the conditions of the contract. It is difficult to see, in these circumstances, how insurance in the joint names of the employer and the contractor can be effected, unless the contractor signs the building contract and then insists that a separate insurance policy is taken out by the employer for the specified work to be undertaken.

Generally the terms of building contracts define the provisions for insurance and the possible events following fire damage in respect of reinstatement. For example, in clause 22A of the JCT Standard Form 1980 Edition, provision is made for insurance to be in the joint names of the employer and the contractor. The contractor is responsible for arranging the policy and paying the premium, and he is instructed to insure for the full reinstatement value of the building together with the cost of professional fees. No specific reference is made to provision for inflation costs during the reinstatement period and, if the sum insured is the same as the contract sum, this could prove inadequate. The situation would become more severe where, for commercial reasons, during a period of economic recession the contractor has entered into a building contract at a loss-making tender price. The full reinstatement figure must be a realistic sum, or any claim, whether for partial or total reinstatement, may be subject to average.

The JCT clause mentioned above details the procedure for repair and defines the method of payment to the contractor for the reinstatement costs. This clause implies that the architect has some responsibility to the insurer, but this could only apply if it is written into the insurance policy.

What is not stated in this clause, and is omitted from most building contracts, is insurance provision relating to loss or expense to the employer being denied the use of the completed building as a result of fire damage. This could be loss of rent, presumably equivalent to the amount of the stated liquidated damages – a consequential loss situation. If proper insurance has not been provided to cover this point, no claim for compensation could be made on the building contract policy.

It is also possible that the compensation to the contractor might not be completely provided solely by complying with insurance requirements in a building contract. He too may have a consequential loss situation in that the building contract may not provide him with adequate protection to meet any loss or expense for an extended contract period arising from delay by fire damage.

This matter has been dealt with under one heading. The need for adequate insurance arises at the time of placing the risk and, as in all insurance, the claim will relate to the wording of the policy and the sum insured. Employers and contractors should carefully review their insurance cover prior to proceeding with a building contract, and consider the effect that fire would have on their affairs if damage occurred during construction.

Value added tax

VAT will apply to most reinstatement building costs and to all professional fees, provided the consultant is registered. The situation with VAT has been explained in detail in previous chapters. At claim stage, the need is to ensure reimbursement from insurers, provided the policy and sum insured have taken this into account, or the status of the parties allow full reimbursement from HM Customs and Excise.

Repair

Whether the fire damage repair is by reinstatement or reconstruction in a manner different from the original, various alternative tendering procedures and contractual arrangements are available. These are dealt with in detail in the Aqua Group book entitled *Tenders and Contracts for Building*.

The professional team should select the best combination to suit the particular circumstances. As an *aide memoire* to this subject the lists below give the principal options available:

Tendering procedures

- Competition
- negotiation with one builder.

Tender documents leading to contract documents

- Bills of quantities
- bills of approximate quantities
- schedule of rates
- specification and drawings
- cost reimbursement
- target cost.

Contractual arrangements

- Firm price
- fluctuating price
- standard forms with or without quantities or with approximate quantities
- fixed fee standard form of prime cost contract
- standard form of agreement for minor building works
- management contract
- design and build contract.

Preferred option

Insurers require reinstatement to be quick and competitive. The preparation of detailed drawings, specifications and tender documents will take time and this is further increased by a month or so needed to obtain competitive tenders, and to assess and recommend the contractor to carry out the work. Mobilisation by the contractor follows and the commencement on site can be several months after the fire damage. This arrangement would not accord with the wishes of many insurers – too much paper up front and too much delay in beginning site operations.

The preferred option is likely to embrace the use of subtrade packages under the contract control of a main contractor.

During the period that the initial emergency work is being carried out, the design team can produce an outline specification, draft general arrangement drawings and an enabling tender document, to define packages to be tendered competitively and let as the work proceeds. The contractual arrangements can be based on a fixed fee standard form of prime cost contract, suitably adapted to operate a document providing the requirement of the contractor to price:

- preliminaries;
- fixed fee;
- provisional sum for materials to be set against invoice costs with opportunity for contractor to add a percentage;
- provisional sum for subtrades to be set against trade packages tendered competitively with opportunity for contractor to add a percentage;
- prime cost sum for nominated subcontractors to be set against packages tendered competitively with opportunity for contractor to add a percentage;
- provisional hours for each trade to be quoted by the contractor in labour rates to be set against time sheets.

The option would require the design team to estimate a reasonable allocation of the cost plan to each of the above elements, so that a tender figure can be submitted.

Although the tender figure would only give a guide to the final cost, each element will at the end have been tendered competitively. In addition, the period from the date of the fire to commencement on site would in most cases have been reduced to less than one month.

Allocation of risk

There are many risks in all building contracts. Economically each risk bears a possible financial loss. Allocation of these risks should therefore reflect who will be responsible for the financial costs.

An insurance policy including a reinstatement memorandum should indemnify the insured against the cost of building reinstatement. All building contracts will be influenced by:

- time in design, procurement and construction
- specification and quality of work
- costs.

Provided the sum insured is adequate and the claim is valid, the principal risks of reinstatement remain with the insurers. The designers must assume the risk of design and the builder must assume the risk of construction and the terms of the building contract. Beyond these points there appears to be no necessity to burden the designers and the builder with additional risks.

Where the extent of the reinstatement is unclear, the financial risk of determining that extent should remain with the insurers. Full tests and investigations should be carried out prior to the building contract on drains, electrical, heating, lifts and other service installations, to determine the extent of the reinstatement. Similarly, site work may be necessary to determine the extent of damage to foundations, structural members and items for demolition. All these hidden risk items should remain as part of the indemnity by the insurers and related to proper and established costs.

It is vital that procedures prior to the insured entering into a building contract for reinstatement are clearly documented and agreed with the appointed loss adjuster. Allocation of risks in the reinstatement should be clearly defined and the method of payment agreed.

Where a building contractor is unable to complete the reinstatement and it becomes necessary to determine the contract, the risk of indemnity still remains with the insurers. For example, should a contractor become bankrupt, the liabilities with subcontractors, merchants and labour will be defined in the building contract or by legislation. The risk of indemnity to complete the reinstatement lies with the insurers provided that the sum insured is adequate.

Subrogation

Most property insurance policies include a subrogation clause, although an express provision is not necessary for the right to arise. Subrogation can be defined as the substitution of one person or thing for another in respect of a right or claim. In particular it applies to the placing of a surety who has paid the debt in place of the creditor and then entitling the surety to seek repayment of that debt in law from the party causing the original liability.

There are many legal cases based on subrogation. Some are relatively simple and need little testing in the courts. One example could be the negligent use of a blowtorch by a professional decorator during redecorations, resulting in fire damage to the insured house or adjoining property. The insured would be reimbursed under the terms of the policy, leaving the insurers by subrogation to seek compensation from the decorator. If the damage was to an adjoining property, then the insurers of that property would meet the reinstatement under the terms of the policy, and by subrogation would seek compensation from the policy holder of the house being redecorated or the decorator or both.

Other cases can be more complicated and more testing in law. Such a case is the action by *Digital Equipment Co. Ltd* (plaintiff: Digital) v. *Hampshire County Council* representing the Hampshire Fire Brigade (defendant: Fire Brigade) [June 1996], Court of Appeal.

An office building in Basingstoke was severely damaged by fire and the owners of the building (Capital & Counties plc) and the leaseholder (Digital) were compensated under the terms of their insurance policies. The insurers, through subrogation, took legal action against the original builders, including specialist subcontractors, maintenance contractors, manufacturers of electrical products, the designers and the fire brigade. Apart from the fire brigade, all matters were settled with the other defendants. The fire brigade continued with their legal defence against the claim of negligence throughout.

In this case, a relatively small fire had commenced as a result of an electrical fault. The building had been evacuated and the automatic sprinkler system was operating and containing the fire. The Fire Brigade responded by sending one appliance, which prompted the leading fire fighter, on arrival, to call for additional appliances. Some 25 minutes after the arrival of the first appliance, the station officer of the fire brigade ordered that the entire sprinkler system should be shut down. Within an hour or so an extensive fire had almost destroyed three blocks, which formed the office development.

During the course of the legal proceedings, the plaintiff put forward the following allegations and issues against the Fire Brigade:

- failed to respond with sufficient appliances;
- failed to inspect and make themselves aware of the automatic sprinkler system installed in the property;
- failed to fight the fire in a professional manner;
- failed to appreciate the consequences of shutting down the sprinkler installation;
- that they (the fire brigade) owed a duty of care to the owner of the property;
- that they had breached that duty of care.

Without detailing the judgment given, the judge stated that the fire brigade had been found negligent in shutting down the sprinkler installation and found in favour of the plaintiff. Hence by subrogation the insurers of the Digital property were able to regain the money that they had paid out pertinent to the insurance policy.

Indemnity for fire brigades and the need to insure

The legal case of *Digital Equipment Co. Ltd* v. *Hampshire County Council* [1996] is included in this book as a good illustration of the application of subrogation. It also shows that a fire brigade can be guilty in the tort of negligence where they themselves have created the danger; they have no legal protection through relevant statutes where self-created civil wrongs occur. The common law of duty of care applies.

However, in another case, the *Church of Jesus Christ of Latter Day*

Saints (Great Britain) v. *Yorkshire Fire and Civil Defence Authority* (1996), the legal action relied on the absence of a proper supply of water. Seven fire hydrants were situated around the Huddersfield chapel but four failed to work and three were either never found or found so late as to be of little use. As a result, water had to be obtained from a mill dam some half a mile away from the chapel and in the ensuing delay the fire, which should have been contained to a classroom, had destroyed the entire chapel.

The plaintiff alleged breach of duty by the fire brigade in that the hydrants had not been regularly inspected, that defects in them had not been observed or repaired and that one hydrant had become hidden by vegetation. The judgment held that at common law the cause for possible future damage was clearly foreseeable, but applying the test in *Caparo Industries plc* v. *Dickman* [1990], it was not fair, just and reasonable to superimpose on the statutory framework a common law duty of care. Furthermore, it was held that it would create the possibility of massive financial claims, which had never before been contemplated and which would be an unreasonable burden on the tax payer.

The conclusion was for the individual to insure the property against fire and it was not for the community to become the insurer.

bgb.

Appendix 1

Example of valuation report for fire insurance purposes

VALUATIONS OF BUILDINGS
based on
NOTIONAL REPLACEMENT COSTS AT DAY ONE
(31 DECEMBER 1997)
for
FIRE INSURANCE PURPOSES
for
INDUSTRIAL, COMMERCIAL AND DOMESTIC PROPERTIES
owned by
ENDURABLE LAMPS LIMITED

The following figures are the valuation costs of reinstating the buildings after fire damage, with materials and in a style similar to that which now exists, including the following items where applicable:

- superstructures and substructures, including ground floor slabs and all internal finishings;
- machine bases cast in with the floor slab;
- partitions of all types including masonry, timber, steel and glazing;
- services and drains below ground floor slabs;
- plumbing and electrical installations;
- sprinkler installations;
- air handling installations;
- heating installations;

- boilers and oil storage restricted to heating and hot water installations;
- passenger and goods lifts;
- name signs;
- light fittings;
- paved external areas;
- landscaping.

Where the existing construction does not appear to comply with the present day building regulations we have allowed in our valuations for the cost of complying with those requirements in any reinstatement. The valuation figures do not allow for any improvements in the reconstruction of the existing building in the event of reinstatement after fire damage, with the exception of the minimum requirements to comply with building regulations. Improvements to comply with planning requirements are excluded from this valuation. No investigations have been made by us that planning approval would or would not be given in the event of serious damage by fire to the properties in this report.

You have advised us that the following items are insured separately through first loss policies or through your contents insurance policy and *no* allowance has been made in the valuations for:

- local authority charges and VAT thereon;
- professional fees and VAT thereon;
- loss of rent;
- cost of alternative accommodation;
- demolitions and debris removal;
- free standing racking;
- display shelving;
- carpets and floor coverings not stuck to the floor slab;
- furniture, stock, plant, machinery and contents;
- personal items in domestic property;
- canteen and kitchen equipment;
- computers and associated equipment;
- telephones and public address installations;
- petrol and oil pumps and storage tanks;
- weighbridges and pits;
- external flood lighting;

- fire fighting equipment;
- compressed air lines;
- burglar alarms;
- gantry cranes;
- open unpaved car or lorry park areas; } insurance not required
- hidden redundant work. }

The valuations make no allowance for VAT, which will be chargeable as laid down in the Finance Acts. You should obtain professional advice on the matter of the VAT charge arising between a builder and yourselves in the event of reconstruction or partial reconstruction after fire damage and any amount you are unable to recover from HM Customs and Excise should be added to our valuations in determining the basis for the sum insured.

The basis of the valuations is building prices current at 31 December 1997. The figures given are the notional replacement costs on that day ('Day One') and *exclude* all provisions in respect of inflation relating to:

- increased costs for the period of planning, tender and site clearance;
- increased costs for the period of reconstruction;
- increased costs for the period of insurance.

During inflationary periods, at no time will the valuation figures given in this report be adequate to cover the cost of total reinstatement. Please note that any inflation arising from the three separate periods stated above will need to be recoverable from the insurers by you under the specific provisions of the policy wording to be recommended by your insurance brokers. In some instances this inflation factor would apply to a total period from 'Day One' of as much as three years and the resultant increased costs in addition to our valuations would be considerable. The recommended allowance for this matter is not less than 20%.

The valuation figures are based on the assumption that, in the event of fire damage, adequate time would be available in order to plan the reconstruction in an economical manner and competitive tenders would be obtained from builders. The effect of any attempt to shorten the reinstatement period to reduce consequential loss claim is excluded from these valuations.

Where properties have multi-occupation comprising a number of self-contained units, the valuation given is for the building as a whole. The sum insured should show no division in respect of individual flats or units. Any loss, whether it be partial or total, should be set against the whole sum insured.

Existing automatic sprinkler installations do not comply with the latest and current edition of The Loss Prevention Council (incorporating FOC). Insurance arranged on the sprinkler installation should include in the policy a special memorandum that additional costs have been included in the sum insured to allow the installations, when reinstated, to comply with the current edition on this subject.

Your insurance policy should include a supplementary clause which will ensure that you are reimbursed with investigation costs of testing the whole of the drainage, plumbing, electrical, heating and other service installations to the buildings in the event of partial material damage by fire, water or other insured risks, and the cost of subsequent partial or total reinstatement.

The reinstatement values must *not* be taken as a guide to the current market value of the properties and reference to this valuation may not be included in any published document without the surveyor's written approval.

The valuations of buildings based on notional replacement costs and qualified as described (excluding all allowances for inflation) as at 31 December 1997 are as follows:

	Day One
Low Lane, Horsforth, Yorkshire	8 748 000
18–36 David Road, London NW10	2 675 000
Arthur House, Goldhawk Road, London W12	1 760 000
Total	**£13 183 000**

Low Lane, Horsforth, Yorkshire (date of survey – October 1997)

Gross internal floor areas:

- Plan refs 1 and 2: 18 910 m^2
- Plan ref. 3: 2 570 m^2
- Plan ref. 4: 61 m^2

Plan refs 1 and 2 Factory, warehouse, workshops, boiler room, laboratories, toilets, canteen, kitchen and air handling plant areas Two storey and plant mezzanine over part of area Brick and PVC covered steel wall cladding; reinforced concrete and steel framing; concrete floors; flat felted and decked roof; air handling and heating installation; sprinklers including storage tank	£6 620 000
Plan ref. 3 Offices Three and four storey and roof housing Brick walls with slate dressings; reinforced concrete framing, floors, staircase and roof; good quality internal and external finishings; brick and concrete external fire escape stairs	£1 930 000
Plan ref. 4 Sprinkler pump housing One storey Brick walls; concrete floors; concrete roof	£36 500
External items Brick retaining walls Railings Suspended concrete terrace Tarmacadam paving Shed	£161 500
Valuation of buildings for fire insurance purposes as described.	£8 748 000

18–36 David Road, London NW10 (date of survey – December 1997)

Gross internal floor areas

- Three-storey offices: 840 m^2
- One-storey warehouse: 1870 m^2
- One-storey offices: 78 m^2

Three-storey office block constructed with brick walls, concrete floors, flat roof, internal partitions, passenger lift, radiator central heating and good quality internal finishings (office block vacant on day of survey except for canteen and warehouse office on ground floor)

One-storey warehouse, show and sales area and offices constructed with brick walls, steel framing, concrete floor, pitched glazed and flat roofs, high level blow heating, hoist, goods lift, loading and unloading bay entrances and having mezzanine flooring to part of the floor area. Rear section refurbished 1997

One-storey offices connecting office block to warehouse

All buildings installed with sprinklers

External pavings, fencing and boundary brick walls

Valuation of buildings for fire insurance purposes as described.

£2 675 000

Arthur House, Goldhawk Road, London W12 (date of survey – November 1997)

Gross internal floor areas

• 2 no. shop units:	$272\,m^2$	(ground floor)
	$272\,m^2$	(basement)
• 1 no. shop unit:	$67\,m^2$	(ground floor)
	$94\,m^2$	(basement)
• Upper floors domestic	$548\,m^2$	

Two-, four- and five-storey building comprising three retail shop units on the ground floor with basement accommodation beneath (two units occupied by Endurable Lamps Ltd and one unit occupied by London Driving School) and the upper floors converted into 12 bed/sit flats and one one-bed flat

The building is constructed with brick walls, timber floors and stairs and lead covered flat roofs

Valuation of buildings for fire insurance purposes as described	£1 760 000

Note: The allocation of the insurance premium should be on the following basis:

2 No. – shop units (Endurable Lamps Ltd)	30.5%
1 No. – shop unit (London Driving School)	10.0%
12 No. – bed/sit flats – 12 at 4.2% each	50.4%
1 No. – one-bed flat	9.1%
	100.0%

Appendix 2

Check list of items likely to be included or excluded from a valuation of a building for fire insurance purposes

Items likely to be included

- Structure, substructure, finishings and services
- air conditioning installations
- allowance for complying with Building Regulations and local authority requirements
- boilers used for process power
- boilers used solely for hot water and heating
- built-in display cabinets
- cold room storage
- electrical installations
- electrical work to heating and lighting the buildings, including electric unit heaters and external lighting
- external brick boundary and garden walls
- external pavings, roads, paths, ramps and all main services under these external surfaces
- external screens and fencing
- extractors and ventilators
- fire alarms
- fitted drinks bars
- floor ducts, channels, pits and covers
- goods lifts
- heating installations
- ice-making machines

- internal and external plumbing and sanitary installations and fittings
- kitchen equipment, fixed and plumbed in
- light fittings
- local authority charges and VAT thereon
- machine bases cast in with the floor slab
- partitions of all types including timber, steel and glazing
- passenger lifts
- planted landscape areas (excluding all work to golf courses)
- pumps
- saunas
- septic tank and other drainage disposal plant including piping
- smoke detectors
- sprinkler installations
- structural mezzanines
- sundry structures not specified
- swimming pools
- tennis courts
- water softening plant.

State in the valuation report the items *included*.

Items likely to be excluded

- VAT
- air conditioning installations directly related to items of plant
- boundary retaining walls
- burglar alarm installations
- canteen and kitchen equipment
- carpets and other fitted floor coverings
- casino tables and equipment
- check-out islands and adding machines
- coffee machines
- cold room storage units, together with motors and extractor or fresh air ventilating fans
- compressed air installations
- computers and associated equipment
- computer game machines
- conveyor belts and rollers

- copper covering to walls and floors in coil houses
- copying machines
- counters, grilles, alarms and demountable partitions
- display shelving
- electric fan heaters
- electric hanging signs
- electric night storage heaters
- electricity board's substation transformers and equipment
- external concrete roads and main services beneath
- external embankments
- external flood lighting
- fascia signs
- fire fighting hand equipment
- flume and process drainage installations
- food packing machinery and equipment
- free-standing racking
- freezer units
- fume extract installations
- furniture, machinery and plant generally
- gantry cranes and crane rails
- golf course landscaping and associated fittings and installations
- hairdressing equipment
- hidden redundant work
- kitchen equipment not plumbed in
- lockers
- lorry and car wash installations and equipment
- main services under external pavings and roads
- oil storage tanks above ground
- open unpaved car or lorry park areas
- personal items in domestic and office property
- petrol and oil pumps and storage tanks
- projecting external signs
- refrigeration plant
- rent as included within the insurance policy
- restaurant tables, chairs, fitted seats and equipment, including hot plates and ovens
- safes
- silos, vats, storage tanks and other containers in respect of the manufacturing process

- snooker tables
- stand-by generators
- steel mezzanines supporting or giving access to items of plant
- steel plate floor covering
- stock
- structures of ovens and other manufacturing plant
- sunbeds
- telephones and public address installations
- temporary buildings
- unfixed gym equipment
- video and camera installations for security
- weighbridges and pits
- window blinds and curtains
- work benches.

State in the valuation report items *excluded*.

Items likely to be covered by the wording of the insurance policy

- Inflation
- loss of rent
- cost of alternative accommodation.

State in the valuation report the items covered by the *wording of the policy*.

Items likely to be covered by a first loss policy where the policy covers many buildings and several locations

- Demolitions and debris removal
- professional fees and VAT thereon
- external hard pavings and other external items.

State in the valuation report the items covered by a separate *first loss policy*.

The items given in this Appendix 2 are to illustrate the considerable range of insurable items that can be at risk. The lists are not exhaustive. Neither is the allocation of the items under the above headings recommended, fixed or accepted in all cases. Appendix 1 has deliberately included and excluded items different from the suggestions in Appendix 2.

Appendix 2 should be taken as a helpful check list to agree instructions by the valuer with clients, brokers and underwriters.

Appendix 3

A fire certificate

By permission from the London Fire & Civil Defence Authority, their uncompleted fire certificate is reproduced as a typical example used in the UK, comprising:

Part A – the fire certificate (uncompleted)
Part B – requirements of means of escape and other fire precautionary arrangements to obtain a certificate
Part C – requirements which ensure proper maintenance of matters in Part B to retain a certificate.

Part A - Fire Certificate for the following premises

Address:
Identification of premises: As shown on certificate plan(s), Number(s)
The use or uses of the premises covered by this fire certificate:
LFCDA Case Number:

It is hereby certified:

1. That the above mentioned premises are provided for use in case of fire with:–
   a) the means of escape;
   b) the means for securing the safe and effective use of the means of escape at all material times;
   c) the means for fighting fire in the building;
   d) the means for giving warning to persons in the building;
   as shown on the plan(s) contained in this certificate (construed with reference to the key to plans and definitions herein), which are hereby specified for the purpose of Section 6 of the above Act.
2. That in relation to factory premises, the explosive or highly flammable materials (if any) detailed in this certificate may be stored or used in the premises.
3. That the requirements of the certificate contained in Part C where appropriate to these premises are **HEREBY IMPOSED** in accordance with the allocation of responsibility on Page 3 of Part A.

Signed: ______________________

Assistant Chief Fire Officer
Commander

Date from which this certificate came into effect: [ ]

NOTE: The following form part of this fire certificate:

Part B – The means of escape + other fire precautionary arrangements provided in the premises (to be read in conjunction with plan(s) and key to plan symbols.)

Part C – Schedule of requirements

Please see page 2 for general information about this certificate

General information about this Fire Certificate

1. Responsibility for compliance

Owners/occupiers have a legal responsibility to ensure that the terms of the fire certificate are complied with. (The allocation of responsibility shown on page 3 defines precisely who is responsible for the requirements of the Certificate.)

Owners or occupiers who come to premises after the date of issue of the certificate should note that under the Fire Precautions Act 1971 it is they, **NOT** previous owners or occupiers, who are responsible for compliance.

2. Keeping of the Certificate

Whilst it remains in force, this certificate, (or this copy if issued to an occupier of a multi occupancy building), must be kept in the premises to which it relates. On a change of ownership/occupier, the certificate should be passed to the new owner(s)/occupier and the Authority informed of the change. Standard form FS/A4/8, a copy of which is included with this certificate, can be used for this purpose.

3. Alterations to the premises

(see also information contained inside rear folder cover)

The Fire Authority must be given prior written notice if it is proposed:–

i) to make material extension of or material structural alteration to the premises; or

ii) to make a material alteration in the internal arrangement of the premises or in the furniture or equipment with which the premises provided.

iii) to begin to keep under, in or on the relevant building where your premises are situated explosive or highly flammable materials of a kind and in a quantity prescribed by the Secretary of State or, in the case of factory premises, to begin to store or use explosive or highly flammable materials or materially to increase the extent of such storage or use.

iv) to alter any automatic firefighting system (ie., sprinklers, drencers, deluge installation), other than as an exchange item under routine maintenance, that has been provided in the building for;

- **(a)** life safety system protecting the means of escape routes.
- **(b)** forming part of a fire engineered solution.
- **(c)** fire control purposes as part of a fire evacuation procedure (ie., shopping malls and phased evacuation in buildings in excess of 30m high).

NOTE:
'Material' is taken to mean any alteration which could render the means of escape and fire precautions inadequate in relation to the use of the premises for which the certificate is issued. The Fire Authority's fire safety officers will be pleased to advise you if you are in any doubt.

4. Penalties

Failure to:–

i) keep any of the matters specified in this certificate in accordance with the specification, or failure to observe any of the requirements in Part C;

ii) keep this certificate or, where appropriate, this copy of it, in the premises;

iii) give notice to the Fire Authority before carrying out proposed alterations to the premises (see Section 3 above).

are all offences under the Fire Precautions Act 1971, for which there are prescribed maximum penalties.

5. Rights to Appeal

The Fire Precautions Act 1971 provides a right of appeal to a magistrates' court against various decisions concerning a fire certificate, but you must make your appeal within 21 days from the date on which the decision is made known to you.

Allocation of responsibility for compliance with requirements

In the case of a building with a single occupier
(single occupancy premises

The occupier for the time being of the premises covered by this fire certificate is hereby notified that he/she is responsible for contravention of any of the requirements.

In the case of a building with more than one occupier
(multiple occupancy premises)

The owner(s) and/or occupier(s) for the time being of the premises covered by this fire certificate are hereby notified that they are responsible for contravention of any of the requirements listed below as being their responsibility.

1) The owner(s) is/are responsible for any contravention of a requirement imposed by this fire certificate.

2) In addition to the owner(s), the occupiers are responsible for any contravention of a requirement imposed by this certificate in the premises occupied by them and in those parts of the relevant building which are treated as forming part of the premises for the purposes of Section 1(8), (e.g. parts of the building used in conjunction with the premises).

Part B - The means of escape and other fire precautionary arrangements provided in these premises

Contents

Key to symbols used on plan(s)

Plan (s)

1. Means of escape routes

are safe and unobstructed routes provided for persons to walk from any point in a building to a place of safety away from the premises without outside assistance.

These comprise:-

a) Unobstructed passageways between the furniture, equipment and fittings within a room or on a floor area to exits from that room or area.

b) Corridors.

c) Internal stairways, walkways, ladders and any other routes.

d) Final Exits leading directly from the building to a place of safety.

e) Exits leading to adjoining premises.

f) External stairways, walkways, ladders, and any other routes.

1.1 Protected Means of Escape routes

A stairway, corridor or lobby with enclosures shown by **thick** black lines on plan is part of a Protected route. There will be a letter within a circle to identify the route.

e.g.

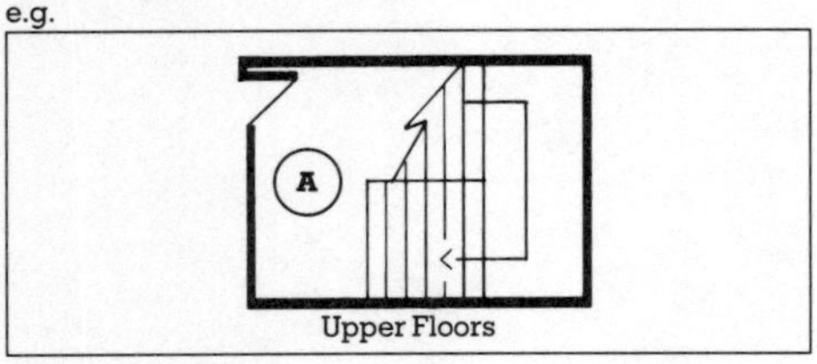

e.g.

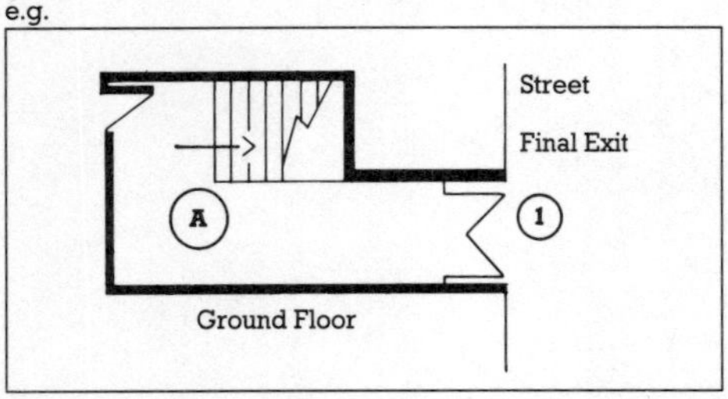

A Protected route has enclosures of FIRE RESISTING construction which ensure that it is always available and safe for use by separating and protecting the route from a fire in the remainder of the building. All doors, walls, partitions, screens and glazing shown in **thick** black (or **red**) lines on the plan are to comply with the definition of FIRE RESISTING on page 4.

1.2 External routes

Routes indicated by a box arrow on plan eg. ⟹ are 'External routes.' There will be a letter within a circle to identify the route.

e.g.

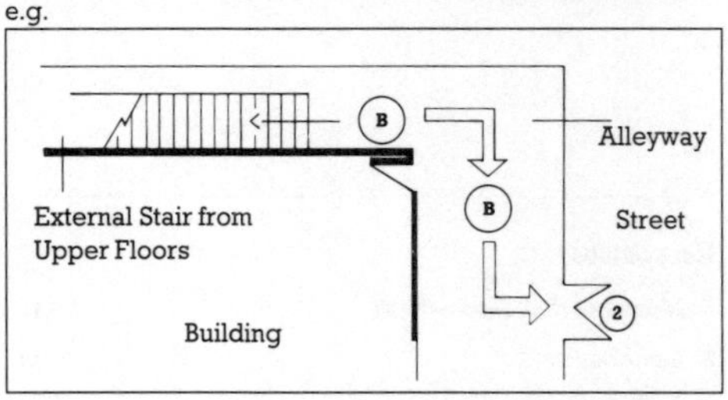

These routes may be by way of a roof, stairway, ladder, open area, courtyard, walkway or adjoining buildings.

External routes are protected from a fire in the building by the adjacent doors, windows and external walls being of FIRE RESISTING construction. All doors, windows and external walls shown in a thick **black** (or **red**) line on plan are to comply with the definition of FIRE RESISTING on page 4.

1.3 Final exit

A Final exit is shown on plan by a number in a circle, eg. ① This indicates the termination of an escape route from the building which gives direct access to a place of safety, such as a street, passageway, walkway or open space and sited to ensure that persons can disperse safely from the vicinity of the building and the effects of fire.

2. General

2.1

Internal and External stairways, landings, elevated walkways, ladders and roof escape routes are provided with handrails and are adequately guarded on any open side.

2.2

Any means of escape route which has no natural or borrowed lighting, and which may be used during the hours of darkness, is provided with artificial lighting of a sufficient standard to enable persons to leave the premises safely.

2.3 Walls and ceiling finishes.

The surface finishes of walls, partitions and ceiling shall accord with the relevant Home Office guide to the Fire Precautions Act 1971, except as otherwise stated on plan.

2.4 Decorative furnishings.

Decorative furnishings fixed on or adjacent to walls, partitions or ceilings within any staircase, corridor or lobby which forms part of any means of escape route shall be of non-combustible, inherently non-flammable or durably flameproofed material as defined by BS 5438 and BS 5867.

3. Definition of Fire Resisting

3.1 Fire Resisting

means that the construction complies with the Building Regulations current at the date of this certificate for resisting the action of fire for a period of not less than thirty minutes.

and that:–

(a) The construction gives effective separation between those parts of the premises on each side of it;

(b) Where the word 'enclosure' is used in connection with this Standard Term, the room, escape route or other space has ceilings and floors of **Fire Resisting** construction except where that part of the enclosure consists of a roof or the lowest floor of the building.

3.2 Fire Resisting Doors

i) The door and frame, when tested in accordance with the British Standard in force at the date of its manufacture, is able to satisfy the criteria for integrity for a period of 30 minutes (unless otherwise indicated in figures adjacent to each door on plan.) It shall also be able to resist the passage of smoke when tested in accordance with the above standard if so indicated by the letter 'S' adjacent to the door on plan.

ii) The door is effectively self-closing (unless it is to a duct, shaft or cupboard when, alternatively, it may be locked shut and permanently marked 'KEEP LOCKED SHUT' in 5mm lettering).

iii) The door is permanently marked 'FIRE DOOR KEEP SHUT' in 5mm lettering unless it is a bedroom door in a hotel or boarding house.

3.3 Fire Resisting Glazing

Where the construction incorporates glazing, the glazing is in a frame fixed shut.

4. Fire Alarm System Specification (Single Stage Type)

4.1 The fire alarm system (which includes any smoke detection system) for this building is to be a single stage type. This means that the whole occupancy of the building is to evacuate on actuation of the fire alarm system. It shall conform to the British Standard specified within the engineer's certificate of compliance, if annexed to this certificate (see 4.9 below), or the British Standard applicable at the time of installation.

4.2 Fire Alarm Call points shall be located as indicated by on the plan.

4.3 The system shall have alarm sounders, eg. bells, hooters, loudspeakers or sirens, in circuit with the call points. The sounders shall be positioned where they will give a clearly audible warning throughout the building and the number and location of sounders necessary to achieve this recorded on the engineers certificate annexed to the fire certificate (see 4.9 below).

4.4 The type of sounder is to be such that the alarm cannot be confused with any other sound and will be heard distinctly throughout the building, above background noise. In the case of premises providing sleeping accommodation, it shall be capable of waking any sleeping persons.

4.5 The power supply is to be taken from the mains supply through a separate switch fuse, independent of any other circuit.

4.6 A secondary source of supply, which is capable of operating the system in the event of mains failure shall be incorporated, i.e. a battery maintained adequately charged by an AC mains operated charger.

NOTE:
The length of time for which the battery shall be capable of operating the system shall be determined by the use of the premises.

4.7 An automatic fire detection system in circuit with the electrical fire alarm system, shall be installed in areas as indicated by (Smoke Detectors) or (Heat Detectors) on plan.

NOTE:
These locations indicate the areas bounded by walls, partitions and doors to be covered by automatic detection. **They do not indicate the position of individual detectors.** This is particularly important in the case of corridors, irregularly shaped areas, and suspended ceiling voids etc.

4.8 An Indicator/Control Panel shall be located as indicated by or or on plan.

4.9 A certificate is to be completed by a competent engineer to indicate that the system complies with the above specifications. This certificate is to be annexed to the Fire Certificate.

Key to Plan Symbols

1. Means of escape and equipment

All doors, walls, partitions, screens and glazing shown in:—

(a) Thick black line, or
(b) Thick black broken line, or
(c) Red

on plan are to comply with the definition of FIRE RESISTING.

BF Doors which are to have only a break fastening which is immediately and easily openable by a person on his/her way out without use of a key eg.

(a) Lever handle covered by a box and thin glass.
(b) Patent catch covered by a glass seal.
(c) Glass bolt.

There must be a notice adjacent to the fastening indicating the method of opening. (minimum 5mm lettering).

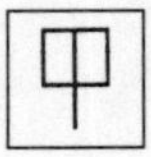

Fire Alarm Call Point with Fire Action Sign adjacent, the contents of which should be based on The Fire Safety and Evacuation Plan for the premises.

SL Door, which may be provided with special fastenings as described on plan.

FFF Doors which are to be free from all fastenings (other than a ball or roller fastening or a lever handled latch). This is to ensure that the door is always available for escape from either side.

Control panel sounders.

Control panel illuminated signals.

Control panel sounders and illuminated signals.

SCA Self closing door which may be held open by an **Electro Magnetic Door Holder.**

PB Doors which are to be provided with a panic bolt. There must be a notice adjacent to the fastening indicating the method of opening (minimum 5mm lettering).

SF Doors which are to have only a simple fastening which is easily and immediately openable by a person on his/her way out without use of a key, eg. barrel bolts, night latch, lever handle, etc.

TP Fire Alarm Telephone point.

VP A panel of clear glass is to be provided in the door or adjacent partition shown on plan. It should be of an appropriate size and in a suitable position which gives the occupant of the inner room early visual warning of fire. This glass should be **Fire Resisting** if the door or partition is shown as such.

S Door assembly which is able to resist the passage of smoke when tested in accordance with the British Standard in force at the date of manufacture.

PR Pressurisation system.

MSE Mechanical Smoke Extraction.

NSE Natural Smoke Extraction.

POC Pressurisation Override Control.

2. Fire Fighting Equipment

Aqueous Film Forming Foam Extinguisher

Water Fire Extinguisher

Fire Blanket in container

Carbon Dioxide Fire Extinguisher (or Halon)

Dry Powder Extinguisher

Foam Fire Extinguisher

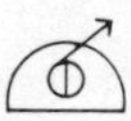

Hydraulic hose reel. A number following this symbol indicates the length of the tubing in metres.

Sand Bucket

Area covered by Fixed Fire Extinguishing installation (i.e. Halon, Co2).

Area covered by Automatic Sprinkler installation.

A number following the symbol indicates the rating of the extinguisher as appropriate.

All fire fighting equipment should be so placed as to be readily available for use. They should be mounted on brackets or suitable shelves in conspicuous positions so that the carrying handles of the larger heavier extinguishers are approximately 1m from the floor, and the handles of the smaller fire extinguishers/base of fire blankets are approximately 1.5m from the floor.

3. Other Codes

Area covered with a system of Escape Lighting which will illuminate the area upon failure of the normal lighting power supply, to a sufficient standard to enable persons to leave the area safely. The escape lighting system should conform to the British Standard current at the time of its installation and a certificate to this effect kept with the fire certificate.

Area covered by Automatic Heat Detectors.

Area covered by Automatic Smoke Detectors.

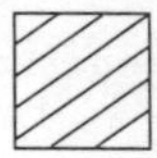

Areas restricted in use: Areas indicated by black hatching on plan are parts of the premises where the means of escape are deemed satisfactory, provided that they are only occupied by persons occasionally in connection with storage or maintenance of fixed domestic services and the building fabrics in those areas.

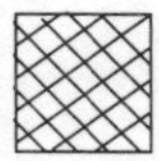

Parts of premises not covered by this certificate.

PR Area covered by Pressurisation system.

MSE Area covered by Mechanical Smoke Extraction.

NSE Area covered by Natural Smoke Extraction.

POC Pressurisation Override Control.

NOTE:
If a symbol indicates an 'area covered', the extent of that area is defined by the walls, partitions and doors shown on plan. The symbol does **not** relate to individual items of equipment. This is particularly important to note in the case of corridors, irregularly shaped areas, suspended ceiling voids etc.

4. Notices

S22	Boxes like this indicate notices. The following codes show the wording to be displayed on the notices.

All notices should conform generally to the British Standard for signs which is current at the date of installation and the minimum size of lettering is as indicated below unless otherwise shown adjacent to the code on plan.

Additional Notes

S22 →	Directional arrow to be provided with the notice in the direction indicated.
S22 S	Notice to be suspended above head height at right angles to the escape route with the graphic symbol on the approach side.
S22 SS	As above but with the graphic symbol on both sides of the notice.
S22 ⧓	Indicates that the notice is internally illuminated.

S22	Fire exit, exit or emergency exit notice graphic symbol	***Clearly visible from furthest point along escape route***
S23	Slide to open (with arrow indicating direction)	***Minimum size of lettering - 25mm***
S20	Fire escape keep clear	
S19	Gangway keep clear	
S17	Secure door open when premises are occupied	
S8	Highly flammable material – keep locked	
S2	No smoking	
S13	Fire door keep shut	***Minimum size of lettering - 5mm***
S14	Fire door keep locked	
S12	Fire Action Sign	

Part C - Schedule of requirements which ensure proper maintenance of the means of escape and other fire precautionary arrangements

Contents

1. Means of Escape, Escape lighting, Fire Fighting Equipment, Alarm Systems, etc.

The following matters specified in this certificate:

a) The means of escape

b) The means for securing the safe and effective use of the means of escape at all material times (e.g. normal and escape lighting, exit signs etc)

c) The means for fighting fire provided in the building

d) The means for giving warning to persons in the building (i.e. the fire alarm system.)

shall be kept in accordance with the specification in this certificate and maintained for use in case of fire properly in accordance with any detailed requirements listed below relating to these matters.

2. Maintenance of Means of Escape Routes

2.1 All means of escape shall be maintained free from obstructions and goods or any materials which constitute a hazard to the route. Stairways, corridors, and lobbies shall not be provided with furniture or equipment unless specified on plan.

2.2 Floors, stairs, ramps and steps in all means of escape routes shall be properly maintained in a safe condition with non-slippery and even surfaces, and with any floor coverings properly secured in position.

2.3 Furniture and other fittings in occupied areas shall always be arranged so as to provide reasonably direct travel from any position in the area to the exit from that area.

2.4 Maintenance and building works shall not be carried out at a time or in a manner which would obstruct or degrade the means of escape from the premises.

2.5 Handrails and guardrails to internal and external stairs, landings, sloping ladders and walkways shall be maintained in a safe and effective condition.

2.6 All external escape routes, including stairs, landings and gangways shall be kept safe for use at all material times, including keeping them clear of snow/ice during inclement weather.

2.7 The external staircases, balconies and gangways which form part of the means of escape from the premises shall be examined by a competent person not more than three years after the date of this Certificate. They shall be examined thereafter at intervals not exceeding three years and all necessary repairs and maintenance work shall be carried out. A written report of every examination, all repairs and maintenance work and a certificate that the structures are in safe and satisfactory condition shall be completed by the competent person and attached to the log book referred to in Requirement 4.4.

2.8 Doors or gates through which persons might have to pass to get out of the premises shall be fastened so that they can be immediately opened by persons on their way out, without the use of a key.

2.9 All self-closing FIRE-RESISTING doors shall be maintained effectively self-closing and free from any means of holding the door in an open position other than an electro magnetic door holder if shown on plan.

2.10 All doors shown on the plan with a symbol indicating that they shall be kept free from fastenings shall be maintained free from all fastenings other than a ball or roller fastening or a lever handled latch.

2.11 Doors shown on plan as opening both ways across escape routes and corridors shall be maintained and shall have vision panels.

2.12 Doors opening in direction of escape shall be so maintained.

2.13 Doors shown on plan as fitted with panic bolts shall not be provided with any other fastenings except for security arrangements if specified on plan.

2.14 Doors shown on plan as having 'Break to Open' fastenings shall be regularly opened at least once each month to ensure that they are in good working order and the test recorded in the log book.

Means of Escape to adjoining premises (if shown on plan)

2.15 Safe and easy access in the event of fire shall be maintained for the occupants of the premises covered by this certificate to enter the adjoining premises specified and reach the street.

NOTE:
The occupier and/or owner shall satisfy himself/herself that he/she has an agreement which will ensure that access will be available whenever necessary, since any obstruction or interruption to the escape route may be an offence under the Act.

3. Lighting

Artificial lighting to ESCAPE ROUTES

3.1 Artificial lighting to ANY MEANS OF ESCAPE ROUTES without natural or borrowed lighting shall be kept alight during the whole of the time the premises are occupied. Other means of escape routes shall have a suitable system of control to enable the artificial lighting to be used when the natural or borrowed lighting provides insufficient illumination to allow the route to be used safely.

Escape lighting

3.2 The escape lighting system shall be maintained in satisfactory working order and be tested and serviced in accordance with the relevant British Standard current at the date of its installation and as recommended by the manufacturers instructions. A copy of the manufacturers recommendations in this respect shall be kept with the log book. A certificate stating that the system complies with the specification contained in Part B is to be kept with the fire certificate in the premises.

The results of all tests, a record of the maintenance, and 6 monthly and 3 yearly test certificates shall be kept in the log book.

4. General

Wall and Ceiling Finishes/Decorative furnishings

4.1 Walls and ceiling finishes and decorative furnishings shall be maintained as specified in Part B of this fire certificate.

Any proposed changes which may increase the fire hazard associated with the above items shall be notified to the Fire Authority.

Further details can be obtained from the local Fire Safety Office.

Flammable substances

4.2 Explosive or highly flammable substances (other than small quantities in daily use or those specified on plan), shall not be kept on the premises. Substances specified on plan shall be kept where stated and in amounts not greater than those stated.

Notices and signs

4.3 All notices required in this Certificate giving an instruction or prohibition shall be complied with. All notices and signs required by this certificate shall be maintained legible and unobstructed in their allotted positions.

Log book

4.4 A log book shall be kept in which there shall be recorded details of tests, examinations and fire drill instruction. The log book shall be kept on the premises and be readily available for inspection as and when required.

Exhibitions, assemblies and public entertainments

4.5 The premises shall not be used for exhibitions, display of goods or as a place of assembly unless the written consent of the Fire Authority has been obtained and the conditions of the consent have been complied with. You are advised that the consent of the local authority may also be required. In this context you should contact the local London Borough Council/City of London Corporation.

5. Fire Fighting Equipment

5.1 All portable fire fighting equipment shall be so placed as to be readily available for use and shall be maintained in satisfactory working order. They shall be examined at least once a year and periodically tested in accordance with the current British Standard and manufacturers instructions by a competent person and the date of such tests shall be clearly marked on the appropriate equipment or on stout tabs securely attached to them.

5.2 Hydraulic hose reels shall be tested once a year to check that they are in working order and produce a satisfactory jet of water, and the test shall be recorded in the log book.

6. Maintenance of Fire Alarm System

Fire Alarm System

6.1 The fire alarm system which is specified in Part B of this certificate, (and automatic fire detection system where installed) shall be maintained in satisfactory working order. The premises shall not be occupied at any time when the fire alarm system is not operational, unless alternative arrangements have been previously agreed by the Fire Authority. An audibility test shall be carried out weekly using a different call point for each successive test to ensure that the fire alarm system and sounders operate satisfactorily and that they can be heard distinctly throughout the building above background noise.

Routine maintenance, tests and checks of the fire alarm system (and automatic fire detection system where installed) shall be carried out in accordance with the British Standard current at the date of its installation and as recommended by the manufacturers. A copy of the manufacturers recommendations in this respect shall be kept with the log book.

The results of all tests and a record of the maintenance shall be kept in the log book. A certificate stating that the system complies with the specification contained in Part B is to be kept with the fire certificate in the premises.

Electro Magnetic Door Holders

6.2 Where Electro Magnetic Door Holders are installed they are to be connected in such a way as to ensure that they are rendered inoperative and that the door closes on the actuation of the fire alarm system.

In the premises with a sleeping risk all doors fitted with electro magnetic door holders must be closed at night.

Each door shall be checked to ensure its correct operation each time the fire alarm is tested. The electro magnetic door holder shall be maintained by a competent person at intervals not exceeding six months and as recommended by the manufacturer. A copy of the manufacturers recommendations in this respect shall be kept with the log book.

Results of tests and maintenance shall be recorded in the log book.

In the event of a door holder or associated smoke detection equipment becoming faulty the relevant door shall be maintained as a self-closing door until the relevant repairs are effected by a comptetent person.

The electro magnetic door holders shall not be altered or adapted in any way likely to affect the efficiency of the fire alarm system.

7. Fire Instruction and Training

7.1 Training

7.1.1 All persons at work and all newly appointed staff shall be instructed and trained to ensure that they understand the fire precautions and the action to be taken in the event of fire. The training shall be appropriate to their responsibilities in an emergency and shall be based on the **FIRE SAFETY AND EVACUATION PLAN** (Section 7.3). A copy of the written instructions shall be given to each person. The instruction and training shall cover:-

(a) the action to be taken on discovery of a fire;

(b) the action to be taken on hearing the fire alarm;

(c) operation of the fire alarm, location of fire alarm call points or internal fire alarm telephones and any indicator panel;

(d) the method of calling the Fire Brigade;

(e) the correct method of operation and the location of fire fighting equipment;

(f) the purpose of fire resisting doors and their location within the premises;

(g) familiarity with all escape routes;

(h) evacuation of the building and location of assembly points.

7.1.2 When parts of the premises are occupied by persons who have a disability which would prevent them from using the means of escape as an able bodied person, efficient arrangements shall be made to ensure their safe evacuation in an emergency. Lifts shall not be used for this purpose unless authorised by the Fire Authority.

7.1.3 All persons at work shall receive verbal instruction not less than once in every period of six months (three months for night duty staff). This shall include evacuation drills (including operation of the Fire Alarms) based on the assumption that, where appropriate, one escape route is not available.

7.1.4 Employees who cannot understand English shall have written instructions provided to them in their own language.

7.1.5 Details of the instruction and training shall be entered in the log book showing (1) date, (2) duration, (3) nature of instruction, (4) name of instructor. A list of names or persons instructed shall be kept with the log book.

7.1.6 All persons on regular duties outside normal working hours shall also be included in the instructions and training.

7.2 Fire Marshals

7.2.1 There shall be at all times, one person nominated as Senior Fire Marshal to be responsible for implementing and co-ordinating the action to be taken in case of fire and sufficient additional persons nominated as Floor Fire Marshals to co-ordinate the actions of all persons. (MINIMUM of one marshal on each floor level.)

7.2.2 Competent persons shall be nominated to take over their responsibilities in their absence.

NOTE:
The implementation of Floor Fire Marshals, described in 7.2.1 above, is necessary to assist the senior fire marshal in evacuation in the event of fire in:

(a) a large building or complex building where numbers of guests, members of the public or staff move frequently from place to place or between floors

(b) where a FIRE SAFETY AND EVACUATION PLAN implements a phased evacuation procédure for the building

(c) a building whose occupants or population are not familiar with the FIRE SAFETY AND EVACUATION PLAN procedures – hotels, theatres, department stores, etc.

7.3 Fire Safety and Evacuation Plan (Single Stage Fire Alarm System)

General Principle

The building has a single stage fire alarm system. This means that the whole occupancy of the building is to evacuate on actuation of the fire alarm system. This is to be co-ordinated by the Senior Fire Marshal and Floor Fire Marshals. The building is not be re-occupied until the Fire Brigade give permission.

Actions of Staff

Persons Discovering the Fire

1. OPERATE FIRE ALARM IMMEDIATELY
2. ATTACK THE FIRE, if possible, with equipment provided but DO NOT take personal risks.

On Hearing the Fire Alarm

General Staff

1. Leave the building immediately and proceed to the assembly point shown below.

   Use **nearest** available exit.

   Do **not** use lifts (except where special arrangements exist for the disabled).

   Do **not** stop to collect belongings.

Floor Fire Marshal(s)

1. Co-ordinate actions of general staff.
2. Ensure evacuation of floor proceeds and is complete by checking all rooms, lavatories etc.
3. Leave building as above.
4. Report details of incident and evacuation complete to Senior Fire Marshal at assembly point.

Senior Fire Marshal

1. Ensure Fire Brigade has been called.
2. Report to assembly point.
3. Record details of incident and evacuation from Floor Fire Marshals.
4. Report details to Fire Brigade on arrival.
5. Assist Fire Brigade if requested.

Special arrangements for disabled (if applicable)

Assembly Point:–

Glossary

Adjuster Acts for the insurer by financial adjustment of indemnity or reinstatement insurance claims, in accordance with the insurance contract and making recommendations for payments from the insurer to the insured. It is usual for the adjuster to attempt to agree with the insured any financial adjustment to a claim. Insurers have the final decision before reference to law.

After-glow Process of continued burning with incandescence, after flame is extinguished.

Assessment Financial calculation by the insured of the insurance claim.

Assessor Acts for the insured by preparing an insurance claim based on indemnity or reinstatement and negotiating a financial settlement with the adjuster acting for the insurer.

Average Relates the sum insured to the value of the material asset at risk and, where the former is less than the latter, reduces the amount of compensation in proportion to that shortfall.

Betterment Financial quantification of the improvement brought about by use of new or alternative materials in place of old, updating of design, or extending the extent of the reconstruction.

Bund (fire wall) Enclosure around plant or tanks containing flammable liquid, formed by the erection of walls or banks above ground level to contain leakage or spillage safely.

Burn To consume or to be consumed by rapid oxidation with the production of heat, usually with incandescence, flame or both.

Charring Formation of a light, friable, mainly carbonaceous residue of wood or other organic matter, resulting from incomplete combustion.

Chemical foam Foam formed by the reaction of alkaline salt solution in the presence of a foam stabilising agent.

Chimney effect Upward thrust of hot gases by convection currents confined within a vertical enclosure.

Combination meter Valve assembly which, when hydrants are opened, automatically permits an increased and separately metered supply of water to be drawn from a consumer's branch water main.

Combustible Capable of burning (opposite of noncombustible).

Combustion See *Burn*.

Compartmentation Division of a building, or large sections of a building, into fire-tight compartments by fire-resisting elements of building construction, in order to contain a fire within the compartment of origin.

Compartment floor/wall Fire-resisting floor/wall used in dividing a building into separate compartments, as defined in the Building Regulations.

Consequential loss A secondary loss arising from the principal incident (e.g. loss of production and business profits following material damage to a factory).

Convection (of heat) Transfer of heat in or by a liquid or gas by the movement of the medium.

Damper, fire resisting Movable closer within a duct, designed to act automatically to prevent the passage of fire. Together with its frame it must be capable of satisfying, for a stated period of time, the criteria of fire resistance.

Dead end Area from which escape is possible in one direction only.

Designated use The purpose for which a building or part of a building is used or intended to be used, the purpose groups being identified in the relevant Building Regulations.

Discharge rate Rate at which a single file of persons can pass through one unit of exit width.

Door, fire resisting Door which, together with its frame, is capable of satisfying, for a stated period of time, the criteria of fire resistance with respect to collapse, flame penetration and excessive temperature rise.

Drencher head, open or closed A water discharge assembly fitted to the pipework of a drencher system, open if the pipework is kept in a dry state, closed and controlled by a detection system if the pipework is kept charged with water.

Drencher system A system of water pipes with drencher heads (open or closed, see above) fitted at suitable intervals and heights, designed so as to protect large surfaces from fire exposure by the discharge of water (usually in heavy quantities) and generally controlled by a fire detection system.

Dry powder Extinguishing medium consisting of finely ground chemicals comprising a main constituent, such as sodium bicarbonate, potassium bicarbonate or ammonium phosphate, and additives to improve its flow, storage and water-repellent characteristics.

Dry riser Vertical pipe installed in a building for fire fighting purposes, fitted with inlet connections at fire brigade access level and landing valves at specified points. It is normally dry but capable of being charged with water, usually by pumping from fire service appliances.

Escape route A route forming part of the means of escape from a point in a building to a final exit.

Evacuation procedure Predetermined plan of action designed to achieve the safe evacuation of the occupants of a building.

Evacuation time Time taken for all occupants of a building or part of a building, on receipt of an evacuation signal, to reach final exit.

Excess Financially the first part of any risk excluded from the policy; a sum stated in the policy which is deductible from any claim.

Explosive range Range of percentage of vapour (gas) concentration by volume in air between the upper and lower limits of flammability.

Exposure hazard Risk of fire spreading from a building to an adjacent separate building, or to another part of the same building, by radiated heat across the intervening space.

Extinguish (fire) To bring a fire to an end by one or more of the following: starvation, inhibition, smothering or cooling; achieved by the reduction of fuel, oxygen and temperature.

Extinguishing system A layout of pipes to distribute a fire fighting agent throughout a building, fitted with nozzles at suitable intervals and heights through which the agent is discharged. May be used with any of the following agents: carbon dioxide (CO_2); dry powder; foam; inert gas.

Fire Ignition: process of combustion characterised by the combination of heat, smoke and flame.

Fire alarm system System of fixed apparatus for giving an audible and/or visible and/or other perceptible warning of fire and which may also initiate other action.

Fire authority The body authorised to act in the enforcement of fire regulations and legislation.

Fire blanket Blanket specifically designed to be used for smothering small fires.

Firebreak Open space separating buildings, stored products or other combustible materials, which is capable of restricting the spread of fire.

Firebreak wall/floor LPC term to define a fire-resisting wall/floor.

Fire certificate Certificate issued by the fire authority certifying that the fire regulations have been complied with for a specific use and stating requirements as to maintenance of fire safety standards.

Fire classification System of classifying fires in terms of the nature of the fuel. The following terms and definitions are taken from BS 4547 (EN 2):

Class A fire fires involving solid materials, usually of an organic nature, in which combustion normally takes place with the formation of glowing embers;

Class B fire fires involving liquids or liquefiable gases;

Class C fire fires involving gases;

Class D fire fires involving metals.

Fire detector Device which gives a signal in response to a change in the ambient condition due to fire. May be of the following types:

gas fire detector detector which responds to the presence of gases produced by combustion or thermal decomposition;

fixed temperature detector

heat detector

heat/fire detector

rate of rise detector detector which responds to an increase in temperature;

infra-red detector radiation fire detector which responds to infra-red radiation;

ionisation chamber smoke fire detector smoke detector which responds when smoke, having entered the detector, causes a change in ionisation currents within the detector;

laser fire detector

laser detector detector which responds to the effect of fire upon the laser beam;

line fire detector detector which responds to fire detected anywhere along its length;

optical smoke fire detector detector having a photo-electric cell which responds when light is absorbed or scattered by smoke particles;

radiation fire detector detector which responds to radiation emitted by a fire.

Fire exit Way out of a room or space to be used by the occupants in the event of fire; also terminal point of an escape route beyond which persons are no longer in danger from fire.

Fire extinguisher Portable item of fire fighting equipment containing a fire extinguishing medium which is expelled by internal pressure or by the action of a self-contained hand pump.

Fire fighting access Approach facilities provided to and/or within a building to enable fire service personnel and equipment to gain access.

Fire lift Lift installed for normal purposes in a building but provided with a priority switch for fire brigade use.

Fire load Total amount of combustion material expressed in heat units, or its equivalent weight of wood.

Fire load density Fire load per unit area.

Fire path Access route from highway for fire fighting or rescue appliances.

Fire point Location where fire fighting equipment is sited, which may also comprise a fire alarm call point and fire instruction notices, the whole being provided and arranged for the use of occupants of premises.

Fire propagation index Comparative measure of the contribution to the growth of a fire of a combustible material, as determined in accordance with BS 476, Part 6.

Fire resistance Ability of an element of building construction to satisfy for a stated period of time some or all of the criteria specified in British Standard, namely resistance to collapse, resistance to flame penetration and resistance to excessive temperature rise on the unexposed face.

Fire resisting Having the ability to withstand the effects of a fire for a specified period of time without loss of its fire-separating or load-bearing function, or both.

Fire resisting separation Separation of parts of a building or structure by means of fire-resisting walls and floors of similar related elements of structure.

Fire retardant Substance or treatment applied to a material to increase its resistance to destruction by fire (also used as adjective).

Fire valve Valve provided on a pipe carrying flammable liquid or gas, designed to cut off the supply automatically or by manual override on the occurrence of a fire.

Fire vent Automatic and/or manually operated vent in the enclosing walls or roof a building for releasing heat and smoke in the event of fire.

Fire wall Wall, screen or separating partition erected in the open air to reduce or avoid risk of radiated heat from or to a building, structure, plant, piece of apparatus or stockpile.

Fireman's switch Switch provided for the use of firemen by means of which high voltage equipment can be isolated from the supply during fire fighting.

Flame Zone of oxidation of gas usually characterised by the liberation of heat and the emission of light.

Flame retardant Substance or treatment applied to a combustible material to decrease its tendency to propagate flame across its surface (also used as adjective).

Flammable (preferred to inflammable) Capacity to burn with a flame (opposite to nonflammable).

Flammable gas detector Device that monitors the atmosphere in a specific area and indicates the concentration of flammable gas present.

Flammability range See *Explosive range*.

Flash-back Rapid movement of flame through a flammable vapour/air mixture from the point where it is ignited to the point where it is generated.

Flash-over Stage in the development of a contained fire at which fire spreads rapidly to give large merged flames throughout the space. As a scientific term 'flash-over' is applicable only to enclosed compartments.

Flash point Lowest temperature at which a liquid gives off sufficient flammable vapour in air to produce a flash (ignition) on the application of a small flame.

Foam compound Mixture of foam-forming materials, such as protein or fluorinated surface active agents, which produce mechanical foam used to smother fires or suppress the emission of gases.

Foam inlet Fixed equipment consisting of an inlet connection, fixed piping and discharge assembly, enabling firemen to introduce foam into an enclosed compartment.

Foam pourer Discharge assembly, either transportable or fixed to the rim of a tank of flammable liquid, which enables foam to be introduced into the tank to form a layer on the surface of the liquid.

Fusible link Soldered connecting link designed to part on melting at a specified temperature.

Hydraulic platform Fire appliance consisting of a platform or cage, supported on hydraulically powered manoeuvrable arms, from which water or foam can be directed onto a fire or with which persons can be rescued.

Ignitability Ability of a material to be ignited by a small flame, as determined in accordance with British Standard.

Ignition temperature Lowest temperature of a substance at which sustained combustion can be initiated.

Incandescence Emission of light by a substance due to its high temperature.

Indemnity cash payment Payment made in settlement of a claim based purely on a financial assessment of the loss and intended to put the insured back, after the loss, into a similar position to that before the incident (the assessment after adjustment by the loss adjuster).

Indemnity value Value of the material asset used as a sum insured – often assessed by depreciating total reinstatement costs by deduction for wear and tear or by reference to market values.

Inerting agent A powder or gas (such as nitrogen or CO_2) which suppresses fire by the physical means of displacing the air in the area of combustion so as to cut off the oxygen supply, or by reducing the temperature below the combustion level.

Inflammable See *Flammable.*

Inhibiting agent Usually a gas which combines inertion with chemical suppression of the combustion process.

Inhibition Process of extinguishing fire by use of an agent that interrupts the chemical reactions in the flame.

Insurance broker A broker acting for the insured and advising on all aspects of insurance, including placing the insured's risks with insurers on the most favourable terms.

Insurance surveyor Acts for the insurers in evaluating the risk on a material asset.

Insured Pays the premium to the insurer and is relieved of the risk.

Insurer Receives the premium from the insured and takes the risk.

Latent heat Heat required to change the state of a substance without causing a change in temperature.

Local authority clause Additional clause attached to, or inserted in, an insurance policy providing for the additional cost in reinstatement of complying with current Building Regulations and other local authority requirements, including planning (note that this prevents the operation of betterment, but the sum insured must be adequate to cover the additional work).

Loss adjuster See *Adjuster*.

Loss Prevention Council UK-based organisation involved in all aspects of loss prevention and control. Supported by the Association of British Insurers and Lloyds. Information is obtainable from: The Loss Prevention Council, Melrose Avenue, Borehamwood, Hertfordshire WD6 2BJ, tel. 0181 207 2345.

Lower limit of flammability Lowest percentage concentration, by volume, of flammable vapour (gas) mixed with air which will burn with a flame.

Material asset The building or item at risk in the insurance policy.

Means of escape Structural means whereby a safe route is provided for persons to travel from any point in a building to a place of safety by their own unaided efforts.

Noncombustible (preferred to incombustible) Not capable of burning.

Nonflammable Not capable of burning with a flame.

Oxidation Combination of oxygen with a substance.

Pressurised escape route Escape route within which the air is maintained at a higher pressure than that in the remainder of the building.

Protected area Area giving an adequate degree of protection from fire in another area and from which there is means of escape.

Protected corridor Corridor forming part or whole of the horizontal component of a protected escape route.

Protected lobby Lobby having an adequate degree of protection and forming part or whole of the horizontal component of an escape route, or affording additional protection to an escape route.

Protected premises Premises or part of premises provided with an automatic system or systems for detecting and/or extinguishing fire, and usually having means of transmitting a fire signal to a remote manned location.

Radiation (of heat) Transfer of heat through a gas or vacuum other than by heating of the intervening space.

Reconstruction Rebuilding after damage incorporating variations from the original design or construction.

Reinstatement Re-establishing the material asset to its former position and restoring it in proper order – 'new for old'.

Reinstatement memorandum Additional clause attached to or inserted in an insurance policy providing for the cost of reinstatement to take place without application of betterment – 'new for old'.

Roof, fire resisting Roof construction which, when subjected to conditions of internal fire, is capable of satisfying for a stated period of time the criteria of fire resistance with respect to collapse and flame penetration.

Roof, resistance to external exposure Ability of a roof deck and covering to resist both penetration by external fire and flame spread over the external surfaces, as determined in accordance with British Standard.

Roof screen Vertical screen fitted internally in the roof a building to divide the roof into bays, so that smoke and hot gases from a fire are contained within the bay of origin.

Safety curtain Vertical fire-resisting screen, operated by gravity with a quick-release device, designed to close the proscenium opening in the event of fire in a theatre or similar place of public assembly.

Salvage Procedures to reduce incidental losses from water, smoke, weather and theft during and following fires.

Salvage equipment Equipment other than fire fighting equipment used to mitigate incidental damage during and following fires.

Schedule of loss Financial assessment representing the notional reinstatement of damage.

Self-ignition temperature Temperature at which a flammable gas/air mixture will ignite without an external source of ignition.

Smoke Visible airborne cloud of fine particles, the product of incomplete combustion.

Smoke outlet Opening, or fire-resisting shaft or duct, provided in a building to act as an outlet for smoke and hot gases produced by an outbreak of fire.

Smouldering Process of combustion without flame but usually with incandescence.

Specific heat Amount of heat required to raise temperature of unit mass by 1°C.

Spontaneous combustion Biological or chemical reaction which produces its own heat, resulting in combustion.

Sprinkler head, open or closed As drencher head, a water discharge assembly fitted to the pipework of a sprinkler system which scatters water automatically in a predetermined pattern over a specified area, open if the pipework is kept in a dry state, closed and automatically opened if the pipework is charged with water.

Sprinkler system A system of pipes with sprinkler heads fitted at suitable intervals and heights, designed so as to control or extinguish fire by the discharge of water in which each of the heads, or groups of heads, opens automatically at a specified temperature or by the operation of a fire detection system.

Sprinkler water accelerator Device fitted to an automatic dry pipe or alternative sprinkler system for accelerating the flow of water into the installation by causing the control valve to operate at a lower pressure differential than the normal setting.

Stairway, enclosed Stairway in a building, physically separated (by walls, partitions, screens, etc.) from the accommodation through which it passes, but not necessarily a protected stairway.

Stairway, protected Stairway having an adequate degree of protection and forming the vertical component of a protected escape route.

Starvation Process of extinguishing fire by limitation or reduction of fuel.

Subrogation The substitution of one person or one thing for another in respect of a right or claim, particularly in placing a surety who has paid a debt in place of the creditor and then entitling the surety to seek repayment of that debt from the party causing the original liability. Most insurance policies include a subrogation clause and there are many court cases on this matter.

Sum insured The maximum sum at risk in the insurance policy.

Surface spread of flame classification Division into classes of combustible building materials according to the rate at which flame spreads over their surfaces, as determined in accordance with British Standard.

Thermocouple Junction of wires of dissimilar metals used for measuring temperature.

Thermostat Automatic control device responsive to changes of temperature.

Turntable ladder Fire appliance embodying a manoeuvrable extending ladder, from which water or foam can be directed onto a fire or with which persons can be rescued.

Valuation for fire insurance purposes Financial assessment of the value of a material asset used as the basis for ascertaining the sum insured.

Ventilated lobby Protected lobby provided with means of ventilation to the open air.

Volatility Readiness with which a substance vaporises.

Wall, fire resisting Wall, either load bearing or nonload bearing, capable of satisfying, for a stated period of time, the criteria of fire resistance with respect to collapse, flame penetration and excessive temperature rise.

Wet riser Vertical pipe installed in a building for fire fighting purposes and permanently charged with water from a pressurised supply, fitted with landing valves at specified points.

Wheeled escape Wheeled extending ladder, usually mounted on a fire appliance from which it can be removed and manoeuvred into position for rescue or fire fighting purposes.

Index